# Lecture Notes in Mathematics

Editors:
A. Dold, Heidelberg
B. Eckmann, Zürich
F. Takens, Groningen

T0253735

Werner Balser

# From Divergent Power Series to Analytic Functions

## Theory and Application of Multisummable Power Series

Springer-Verlag

Berlin Heidelberg New York
London Paris Tokyo
Hong Kong Barcelona
Budapest

Author

Werner Balser
Mathematik V
Universität Ulm
D-89069 Ulm, Germany

Mathematics Subject Classification (1991): 34E05, 34A25, 34A34, 30B10, 30E15, 44A10


ISBN 3-540-58268-1 Springer-Verlag Berlin Heidelberg New York
ISBN 0-387-58268-1 Springer-Verlag New York Berlin Heidelberg


CIP-Data applied for

© Springer-Verlag Berlin Heidelberg 1994
Printed in Germany

Typesetting: Camera ready by author
SPIN: 10130205        46/3140-543210 - Printed on acid-free paper

# Preface

Since the second half of the last century, asymptotic expansions have been an important and very successful tool to understand the structure of solutions of ordinary and partial differential (or difference) equations. The by now classical part of this theory has been presented in many books on differential equations in the complex plane or related topics, by such distinguished authors as Wolfgang Wasow [Wa], Yasutaka Sibuya [Si], and many others. In my opinion, the most important result in this context is (in Wasow's terminology) the Main Asymptotic Existence Theorem: it states that to every formal solution of a differential equation, and every sector (in the complex plane) of sufficiently small opening, one can find a solution of the equation having the formal one as its asymptotic expansion. This solution, in general, is not uniquely determined, and the proofs given for this theorem (in various degrees of generality) do not provide a truly efficient way to compute such a solution, say, in terms of the formal solution. In fact, to prove this result, even for linear, but in particular non-linear equations, and to determine sharp bounds for the opening of the sector (or more generally, determine size and location of all sectors for which the theorem holds, for a given equation with "generic Stokes phenomenon") is not an easy task and has kept researchers busy until very recently; see, e.g., Ramis and Sibuya's paper on Hukuhara domains [RS 1] of 1989, or Wolfgang Jurkat's discussion of Asymptotic Sectors [Ju 1].

In the general theory of asymptotic expansions, the analogue to the Main Asymptotic Existence Theorem is usually called Ritt's Theorem, and is much easier to prove: Given any formal power series and any sector of arbitrary (but finite) opening (on the Riemann surface of the Logarithm), there exists a function, analytic in this sector and having the formal power series as its asymptotic expansion. This function is never uniquely determined — not even when the power series converges. To overcome this non-uniqueness, G.N. Watson [Wt 1/2] in 1911/12, and F. Nevanlinna [Ne] in 1918, introduced a special kind of asymptotic expansions, now commonly called *of Gevrey order* $k > 0$. These have the property that the analogue to Ritt's Theorem holds for sectors of opening up to $\pi/k$, in which cases the function again is not uniquely determined. If the opening is larger than $\pi/k$, however, a function which has a given formal power series as expansion of Gevrey order $k > 0$ may not exist, *but if it does, then it is uniquely determined*. In case of existence, the function can be represented as Laplace Transform of another function, which is analytic at the origin, and whose power series expansion is explicitly given in terms of the formal power series.

This achievement in the general theory of asymptotic expansions obviously escaped the attention of specialists for differential equations in the complex domain for quite some time: In a series of papers [Ho 1-3], J. Horn showed for linear systems of ODE, if the leading term of the coefficient matrix (at a singularity of second kind) has all distinct eigenvalues, and if the sector is large enough, then one has uniqueness in the Main Asymptotic Existence Theorem, and the function can be represented as a Laplace integral, or equivalently, in terms of (inverse) factorial series; however, he did not relate his observations to the general results of Watson and Nevanlinna. Later, Trjitzinsky [Tr] and Turrittin [Tu] treated somewhat more general situations, and they also pointed out the limitation of this approach to special cases.

In 1978/80, J.-P. Ramis [Ra 1/2] introduced his notion of $k$-summability of formal power series, which may best be interpreted as a formalization of the ideas of Watson and Nevanlinna. Applying this to linear systems of (meromorphic) ODE, he proved that every formal (matrix) solution to every such equation can be factored into a finite (matrix-) product of power series (times some explicit functions), so that each factor is $k$-summable, with $k$ depending upon the factor. (In my treatment of *first level formal solutions*, [Ba 3-6] and [BJL], I had, more or less by accident, independently obtained the same result.) This factorization of formal solutions is not truly effective, so that this result did not really give a way to compute the resulting (matrix) function from the formal series.

More recently, J. Ecalle [Ec 1/2] presented a way to achieve this computation, introducing his definition of *multisummability*. In a way, his method differs from Ramis' definition of $k$-summability by cleverly enlarging the class of functions to which Laplace Transform, in some weak form, can be applied. He stated without proofs a large number of results concerning properties and applications of multisummability to formal solutions of (non-linear) systems of ODE. Based upon the described factorization of formal solutions of *linear* equations, it was more or less evident that multisummability applied to all formal solutions of linear equations. However, in the non-linear situation, the first complete proof for this was only very recently given by B.L.J. Braaksma [Br 1]. In which form this result carries over to formal solutions of difference, or other functional equations, is still an open problem upon which much work is done at present. Other directions of activity are the analysis of the Stokes phenomenon for (non-linear) systems, based on the theory of multisummability. Again, Ecalle has done some pioneering work in this direction, but has not given detailed proofs.

As is common in a rapidly growing field, it is difficult for a newcomer to appreciate the results achieved, because in the research papers and monographs at hand, every author chooses his/her own notations and has his/her own ideas of what is elementary or needs to be proved. Concerning notation, I do the same in this text, but at least I am consistent throughout the book, and I have included all proofs — or stated parts of them in form of exercises which I feel readers with some background in Complex Variables should be able to do, if they so wish. In winter semester of 1991/92, I taught a course in Ulm on multisummability. The material covered in this course has become the nucleus of this text, but was considerably expanded and, in some cases, presented in a more elegant form.

I am grateful to Y. Sibuya and B.L.J. Braaksma, who introduced me to the beautiful theory of multisummability in a joint seminar at the University of Minneapolis, in Spring of 1990, and I pray that they will consider me a good student. I also owe thanks to my student Andreas Beck, who went over the proofs and did all exercises, and to Sabine Lebhart for carefully typing the manuscript. Finally, I wish to apologize to my wife Christel, for spending parts of our vacation on a Dutch island on writing this text.


Ulm, 1994                                                          Werner Balser

# Contents

# Chapter 1

# Asymptotic Power Series

This chapter is to set the framework for the remaining ones: We define the notions of *asymptotic expansions*, and in particular, *Gevrey asymptotics*, and we show their main properties. Despite of the fact that we frequently speak of *differential algebras*, a reader is not required to know more than their definition: For our purpose, a differential algebra $A$ is an algebra (over the field of complex numbers), together with a linear mapping $d$ of $A$ into itself which obeys the product rule, i.e. for every $a_1, a_2 \in A$ we have

$$d(a_1 a_2) = d(a_1)a_2 + a_1 d(a_2) .$$

## 1.1 Sectors

Throughout this text, we will deal with analytic functions which generally have a branch point at the origin. Therefore, it is convenient to think of these functions as defined in sectorial regions on the Riemann surface of the (natural) Logarithm. Consequently, complex numbers $z = re^{i\varphi}$ $(r > 0)$ will *not* be the same once their arguments $\varphi$ differ by integer multiples of $2\pi$. Strictly speaking, instead of complex numbers we deal with *pairs* $(r, \varphi)$, but there is little risk of confusion in writing $re^{i\varphi}$ instead of $(r, \varphi)$.

A *sector* (on the Riemann surface of the Logarithm) is defined to be a set of the form

$$S = S(d, \alpha, \rho) = \left\{ z = re^{i\varphi} \mid 0 < r < \rho, \, d - \alpha/2 < \varphi < d + \alpha/2 \right\} ;$$

where $d$ is an arbitrary real number, $\alpha$ is a positive real, and $\rho$ either is a positive real number or $+\infty$. We shall refer to $d$, resp. $\alpha$, resp. $\rho$, as *the bisecting direction*, resp. *the opening*, resp. *the radius* of $S$. In particular, if $\rho = +\infty$, resp. $\rho < +\infty$, we will speak of $S$ having *infinite*, resp. *finite, radius*. It should be kept in mind that we do *not* consider sectors of infinite opening, nor an empty sector. If we write $S(d, \alpha, \rho)$, then it shall go without saying that $d, \alpha, \rho$ are as above. In case $\rho = +\infty$, we mostly write $S(d, \alpha)$ instead of $S(d, \alpha, +\infty)$. A *closed sector* is a set of the form

$$\overline{S} = \overline{S}(d, \alpha, \rho) = \left\{ z = re^{i\varphi} \mid 0 < r \leq \rho, \, d - \alpha/2 \leq \varphi \leq d + \alpha/2 \right\} ,$$

with $d$ and $\alpha$ as before, but $\rho$ a positive real number (i.e. never equal to $+\infty$). Hence closed sectors always are of finite radius, and they never contain the origin.

## 1.2   Analytic Functions in Sectors

Let $S$ be a given sector, and let $f$ be a function analytic in $S$ (hence $f$ may be multi-valued if $S$ has opening larger than $2\pi$). We say that $f$ *is bounded at the origin*, if for every closed subsector $\overline{S_1}$ of $S$ there exists a positive real constant $c$ (depending upon $\overline{S_1}$) such that

$$|f(z)| \leq c \quad \text{for every} \quad z \in \overline{S_1} .$$

If a complex constant, denoted by $f(0)$, exists such that

$$f(0) = \lim_{\substack{z \to 0 \\ z \in S}} f(z) ,$$

*uniformly in every closed subsector*, we say that $f$ *is continuous at the origin*.

If $S$ has opening more than $2\pi$, and $f$ is analytic in $S$, we say that $f$ is *single-valued*, provided that

$$f(z) = f(ze^{2\pi i}) \quad \text{whenever} \quad z, ze^{2\pi i} \in S .$$

We say that $f$ (analytic in some sector $S$) is *analytic at the origin*, if $f$ can be analytically continued to a sector $\tilde{S}$ of opening more than $2\pi$, and if $f$, moreover, is single-valued and bounded at the origin (in $\tilde{S}$); a well-known result on removable singularities then implies that $f$ has a convergent power series expansion about the origin.

Let $S = S(d, \alpha)$ be a sector of infinite radius, and let $f$ be analytic in $S$, or at least analytic for all $z \in S$ with $|z| > \rho$. Suppose that $k > 0$ exists such that the following holds true:

*To every $\varphi$ with $|d - \varphi| < \alpha/2$ and every $r_0 > \rho$ there exist $\varepsilon, c_1, c_2 > 0$ such that for every $z = re^{i\tau}$ with $r \geq r_0$, $|\varphi - \tau| < \varepsilon$,*

$$|f(z)| \leq c_1 \exp\{c_2 |z|^k\} .$$

Then we shall say that $f$ *is of exponential size at most $k$* (in $S$). This notion compares to that of (exponential) order as follows: If $f$ is of exponential size at most $k$ (in $S$), then it either is of order less than $k$, or of order equal to $k$ and of finite type, and vice versa (see, e.g., [Bo] for the definition of order and type, and formulas relating both to the coefficients of an entire function).

**Example:**   Mittag-Leffler's function

$$E_\alpha(z) = \sum_{n=0}^{\infty} z^n / \Gamma(1 + \alpha n), \quad \alpha > 0 ,$$

is an entire function of exponential order $k = 1/\alpha$ and finite type, hence is of exponential size (at most) $k$ in every sector of infinite radius. More generally, if $(f_n)_{n=0}^{\infty}$ is a sequence of complex numbers such that for some $c > 0$

$$|f_n| \leq c^n, \quad n \geq 0,$$

then

$$f(z) = \sum_{n=0}^{\infty} f_n z^n / \Gamma(1 + n/k)$$

is bounded by $E_{1/k}(c|z|)$, and therefore $f(z)$ is of exponential size at most $k$ in every sector of infinite radius.

**Exercises.**

1. Show that if an analytic function $f$ is of exponential size at most $k > 0$ in $S(d, \alpha)$, then the constants $c_1, c_2$ in the above estimate can be chosen independent of $\varphi$, provided $\varphi$ is restricted to a closed subinterval of

$$(d - \alpha/2, d + \alpha/2) .$$

For the following exercises, define

$$g(z) = \int_0^{\infty} e^{zt} t^{-t} dt = \int_0^{\infty} \exp[t(z - \log t)] dt$$

(integrating along the positive real axis); compare also [Nm].

2. Show that $g$ is an entire function and compute its power series expansion.

3. For $\operatorname{Im} z = \frac{\pi}{2} + c$, $c > 0$, show that Cauchy's theorem allows to replace integration along the real axis by integration along the positive imaginary axis. Use this to show (for these $z$)

$$|g(z)| \leq \frac{1}{c} .$$

Prove a similar estimate for $\operatorname{Im} z = -(\frac{\pi}{2} + c)$.

4. For every sector $S$ of infinite radius, not containing the positive real axis, show that $g(z)$ is of exponential size zero in $S$.

5. Use Phragmen-Lindelöf's theorem (see [SG]) or a direct lower estimate of $g(x)$ for $x > 0$, to show that in sectors $S$ including the positive real axis, $g(z)$ cannot be of finite exponential size, hence $g$ is of infinite order.

6. For $c$ as above, let

$$f(z) = g(z + \frac{\pi}{2} + c) .$$

Show that $f(z)$ remains bounded along every ray $\arg z = \varphi$ (for $|z| \to \infty$). Why does this not imply that $f$ is of exponential order zero in every sector $S$?

## 1.3   Formal Power Series

Given a sequence $(f_n)_{n=0}^\infty$ of complex numbers, the formal object

$$\hat{f}(z) = \sum_0^\infty f_n z^n$$

is called a *formal power series* (in $z$). The set of all such formal power series is denoted by

$$\mathbb{C}[[z]] \ .$$

We say that $\hat{f}$ *converges*, or *is convergent*, if $\rho > 0$ exists so that the power series converges for all $z$ with $|z| < \rho$, defining a function $f(z)$, analytic in a neighbourhood of the origin. We shall call $f$ *the sum* of $\hat{f}$ (whenever $\hat{f}$ converges), and we write

$$f = \mathcal{S}\hat{f} \ .$$

The set of all convergent (formal) power series will be denoted by

$$\mathbb{C}\{z\} \ .$$

If $\hat{f}(z) = \sum f_n z^n$ is a formal power series so that for some positive $C, K$, and $k$ we have

$$|f_n| \leq C K^n \Gamma(1 + n/k) \quad \text{for every} \quad n \geq 0 \ ,$$

then we say that $\hat{f}$ is a formal power series *of Gevrey order* $k^{-1}$, and we write

$$\mathbb{C}[[z]]_{1/k}$$

for the set of all such formal power series. It is easily seen (compare the Exercises below) that $\mathbb{C}[[z]]_{1/k}$, under natural operations, forms a differential algebra.

**Exercises.**   For $\hat{f}(z) = \sum f_n z^n$, $\hat{g}(z) = \sum g_n z^n$, and $\alpha \in \mathbb{C}$, define

$$(\hat{f} + \hat{g})(z) = \sum_0^\infty (f_n + g_n) z^n \ , \qquad (\hat{f}\hat{g})(z) = \sum_0^\infty z^n \sum_{m=0}^n f_{n-m} g_m \ ,$$

$$(\alpha\hat{f})(z) = \sum_0^\infty (\alpha f_n) z^n \ , \qquad \hat{f}'(z) = \sum_0^\infty (n+1) f_{n+1} z^n \ .$$

1. Show that $\mathbb{C}[[z]]$, with respect to addition and multiplication with scalars, as defined above, is a vector space over $\mathbb{C}$.

2. Show that $\mathbb{C}[[z]]$, with respect to multiplication of power series, as defined above, is a commutative algebra over $\mathbb{C}$.

3. Show that $\mathbb{C}[[z]]$, with respect to derivation as defined above, is a differential algebra over $\mathbb{C}$ (i.e. show that the map $\hat{f} \mapsto \hat{f}'$ is $\mathbb{C}$-linear and obeys the product rule $(\hat{f}\hat{g})' = \hat{f}'\hat{g} + \hat{f}\hat{g}'$).

4. Show that $\hat{e}$, the power series whose coefficients are all zero except for the constant term which equals one, acts as the unit element of $\mathbb{C}[[z]]$.

5. Show that the invertible elements of $\mathbb{C}[[z]]$, i.e. those $\hat{f}$ to which $\hat{g}$ exists such that $\hat{f}\hat{g} = \hat{e}$, are exactly those whose constant term is non-zero.

6. For arbitrary $k > 0$, show that $\mathbb{C}[[z]]_{1/k}$ again is a differential algebra over $\mathbb{C}$ (with respect to the same operations as above).

   *Hint:* Use the Beta Integral

   $$\frac{\Gamma(\alpha)\Gamma(\beta)}{\Gamma(\alpha+\beta)} = \int_0^1 (1-x)^{\alpha-1} x^{\beta-1} dx , \quad \text{Re}\,\alpha > 0, \quad \text{Re}\,\beta > 0 ,$$

   to show

   $$\sum_{m=0}^{n} \Gamma(1 + \tfrac{n-m}{k})\Gamma(1 + \tfrac{m}{k}) \le (1+n)(1+n/k)\Gamma(1+n/k) .$$

   From this estimate, derive that $\mathbb{C}[[z]]_{1/k}$ is closed with respect to multiplication. Moreover, use Stirling's Formula to show

   $$\frac{\Gamma(1 + \frac{n+1}{k})}{\Gamma(1 + \frac{n}{k})(n/k)^{1/k}} \longrightarrow 1 \quad (n \to \infty) ,$$

   and from this, derive that $\mathbb{C}[[z]]_{1/k}$ is closed with respect to derivation.

7. For arbitrary $k > 0$, show $\hat{f} \in \mathbb{C}[[z]]_{1/k}$ invertible (in $\mathbb{C}[[z]]_{1/k}$) iff it is invertible in $\mathbb{C}[[z]]$, i.e. iff its constant term does not vanish.

   *Hint:* For

   $$\hat{f}(z) = 1 + z\hat{h}(z), \qquad \hat{h}(z) = \sum_{m=0}^{\infty} h_m z^m ,$$

   $$\hat{y}(z) = 1 + z\hat{x}(z), \qquad \hat{x}(z) = \sum_{m=0}^{\infty} x_m z^m ,$$

   show that $\hat{f}, \hat{y} \in \mathbb{C}[[z]]_{1/k}$ iff the series

   $$\sum_{m=0}^{\infty} h_m z^{m/k} / \Gamma(1 + m/k) \quad \text{and} \quad \sum_{m=0}^{\infty} x_m z^{m/k} / \Gamma(1 + m/k)$$

   converge for $|z| < \rho$, with sufficiently small $\rho > 0$, defining functions $h(z)$, resp. $x(z)$, analytic in the variable $z^{1/k}$. Moreover, show that

   $$\hat{f} \cdot \hat{y} = 1$$

is equivalent to the Volterra integral equation

$$x(z) + h(z) + \int_0^z k(z - t)x(t)dt \equiv 0 \,,$$

with

$$k(z) = \sum_{m=0}^{\infty} h_m z^{(m+1)/k-1}/\Gamma((m+1)/k) \,,$$

and use the theory of these equations to show that a (unique) solution $x(z)$ exists which has a convergent expansion of the above form, showing $\hat{y} \in \mathbb{C}[[z]]_{1/k}$ (for such $\hat{f}$ with constant term 1, but this is no restriction).

8. For arbitrary $k > 0$, show that for $\hat{f} = \sum f_n z^n \in \mathbb{C}[[z]]_{1/k}$ with $f_0 = 0$ we have

$$z^{-1}\hat{f}(z) := \sum_0^{\infty} f_{n+1}z^n \in \mathbb{C}[[z]]_{1/k} \,.$$

9. If we interprete $\mathbb{C}[[z]]_{1/k}$ for $k = \infty$ according to the convention $1/\infty = 0$, show that

$$\mathbb{C}[[z]]_{1/\infty} = \mathbb{C}\{z\} \,.$$

Check that the statements of Ex. 6–8 hold true for $k = \infty$.

## 1.4   Asymptotic Expansions

Given a function $f$, analytic in some sector $S$, and a formal power series $\hat{f}(z) = \sum f_n z^n \in \mathbb{C}[[z]]$, one says that $f(z)$ *asymptotically equals* $\hat{f}(z)$, as $z \to 0$ in $S$, or: $\hat{f}(z)$ *is the asymptotic expansion of* $f(z)$ in $S$, iff to every non-negative integer $N$ and every closed subsector $\overline{S_1}$ of $S$ there exists $C = C(N, \overline{S_1}) > 0$ such that for $z \in \overline{S_1}$

$$|z|^{-N} \left| f(z) - \sum_{n=0}^{N-1} f_n z^n \right| \leq C \,;$$

in other words iff the functions

$$r_f(z, N) := z^{-N}\left(f(z) - \sum_{n=0}^{N-1} f_n z^n\right)$$

are bounded at the origin, for every $N \geq 0$. If this is so, we write for short

$$f(z) \cong \hat{f}(z) \quad \text{in} \quad S \,,$$

and whenever we do, it will go without saying that $S$ is a sector, $f$ is analytic in $S$, and $\hat{f}$ is a formal power series.

For this type of asymptotics we refer the reader to standard texts as [Wa], [CL]. For our purposes, we only require the following

**Proposition 1.** a) *Given a sector $S$ and a function $f$, analytic in $S$ and*

$$f(z) \cong \hat{f}(z) \quad in \quad S$$

*for some $\hat{f} = \sum f_n z^n \in \mathbb{C}[[z]]$, the functions $r_f(z, N)$ (as above) are all continuous at the origin, and*

$$\lim_{\substack{z \to 0 \\ z \in S}} r_f(z, N) = f_N \quad (N \geq 0) .$$

b) *Under the same assumptions as in a), suppose that the opening of $S$ is larger than $2\pi$, and that*

$$f(ze^{2\pi i}) = f(z) \quad whenever \quad z, ze^{2\pi i} \in S .$$

*Then $f$ is analytic at the origin, and $\hat{f}$ converges and coincides with the power series expansion of $f$ at the origin.*

**Proof.** a) Observe

$$r_f(z, N+1) = z^{-1}\left(r_f(z, N) - f_N\right) ,$$

hence $r_f(z, N+1)$ bounded at the origin implies

$$\lim_{\substack{z \to 0 \\ z \in S}} \left(r_f(z, N) - f_N\right) = 0 .$$

b) Under our assumptions, $f(z)$ is a single-valued analytic function in a punctured disc around the origin, and remains bounded as $z \to 0$. Hence the origin is a removable singularity of $f$, i.e. $f(z)$ can be expanded into its power series about the origin. It follows right from the definition that the power series expansion is, at the same time, an asymptotic expansion, and from a) we conclude that an asymptotic expansion is uniquely determined by $f(z)$. This proves $\hat{f}(z)$ to converge and be the power series expansion for $f(z)$. □

Let $A(S)$ be the set of all functions $f(z)$, analytic in the sector $S$ and having an asymptotic expansion $\hat{f}(z)$. In view of Proposition 1 a), to every $f(z) \in A(S)$ there is *precisely one* $\hat{f} \in \mathbb{C}[[z]]$ such that $f(z) \cong \hat{f}(z)$ in $S$. Therefore, we have a mapping

$$J: \; A(S) \quad \longrightarrow \quad \mathbb{C}[[z]]$$
$$f(z) \quad \longmapsto \quad \hat{f}(z) = (Jf)(z) ,$$

mapping each $f$ to its asymptotic expansion. Standard results on asymptotics (see [Wa], [CL]) show that $A(S)$, under the natural operations, is a differential algebra, and $J$ is a homomorphism between the two differential algebras $A(S)$ and $\mathbb{C}[[z]]$. Moreover, Ritt's Theorem implies that $J$ is surjective. However, $J$ is not injective — even if we consider sectors of large opening. In the next section we are going to study another type of asymptotic expansions which are better suited to our purposes, since it will turn out that the corresponding map $J$, for sectors of sufficiently large opening, is injective (however, not surjective).

**Exercise.**

Suppose $f$ is analytic in a sector $S$, and for some $\hat{f} \in \mathbb{C}[[z]]$

$$f(z) \cong \hat{f}(z) \quad \text{in} \quad S .$$

If $p$ is a natural number and $\tilde{S}$ is such that $z \in \tilde{S} \iff z^p \in S$, show that $\tilde{f}(z) = f(z^p)$ is analytic in $\tilde{S}$, and

$$\tilde{f}(z) \cong \hat{f}(z^p) \quad \text{in} \quad \tilde{S} .$$

## 1.5  Gevrey Asymptotics

Given $k > 0$, we say that a function $f$, analytic in a sector $S$, *asymptotically equals* $\hat{f}(z) = \sum f_n z^n \in \mathbb{C}[[z]]$ *of order* $k$, or: $\hat{f}$ *is the asymptotic expansion of order* $k$ *of* $f$, iff to every closed subsector $\overline{S_1}$ of $S$ there exist $C, K > 0$ such that for every non-negative integer $N$ and every $z \in \overline{S_1}$

$$|r_f(z, N)| \le C K^N \Gamma(1 + N/k) .$$

**Remark.** Our definition of expansions of order $k$ differs from the classical one: in most papers an expansion of order $k$ would be named *of (Gevrey) order* $k^{-1}$. Consequently, the symbols $A_k(S)$, $A_k^{(0)}(S)$, which will be introduced below, differ accordingly from similar symbols in the literature. This may be misleading for some experts, but should be easier to memorize for students.

If this is so, we write for short

$$f(z) \cong_k \hat{f}(z) \quad \text{in} \quad S .$$

Observe that $f(z) \cong_k \hat{f}(z)$ in $S$ implies $f(z) \cong \hat{f}(z)$ in $S$ (in the previous sense), hence Proposition 1 a) implies that if $f(z) \cong_k \hat{f}(z)$ in $S$, then $\hat{f}(z) \in \mathbb{C}[[z]]_{1/k}$. For $\tilde{k} < k$, Stirling's Formula implies

$$\frac{\Gamma(1 + N/k)}{\Gamma(1 + N/\tilde{k})} \longrightarrow 0 \quad (N \to \infty) ,$$

hence $f(z) \cong_k \hat{f}(z)$ in $S$ implies $f(z) \cong_{\tilde{k}} \hat{f}(z)$ in $S$.

Let $A_k(S)$ denote the set of all $f$, analytic in $S$ and having an asymptotic expansion of order $k$. The following results prove that $A_k(S)$ is again a differential algebra, and $J : A_k(S) \to \mathbb{C}[[z]]_{1/k}$ is a homomorphism.

**Theorem 1.** *Given $k > 0$ and a sector $S$, suppose that*

$$f(z) \quad \cong_k \quad \hat{f}(z) \quad in \quad S ,$$

$$g(z) \quad \cong_k \quad \hat{g}(z) \quad in \quad S .$$

*Then*

$$f(z) + g(z) \quad \cong_k \quad \hat{f}(z) + \hat{g}(z) \quad in \quad S ,$$

$$f(z)g(z) \quad \cong_k \quad \hat{f}(z)\hat{g}(z) \quad in \quad S .$$

**Proof.** Given a closed subsector $\overline{S_1}$ of $S$, there exist $C, K > 0$ such that for every $N$ and $z \in \overline{S_1}$

$$|r_f(z, N)| \le CK^N\Gamma(1 + N/k), \quad |r_g(z, N)| \le CK^N\Gamma(1 + N/k).$$

From

$$|r_{f+g}(z, N) \le |r_f(z, N)| + |r_g(z, N)|$$

we immediately conclude

$$f(z) + g(z) \cong_k \hat{f}(z) + \hat{g}(z) \quad \text{in} \quad S.$$

Moreover,

$$|r_{fg}(z, N)| \le |f(z)| \, |r_g(z, N)| + \sum_{m=0}^{N-1} |g_m| \, |r_f(z, N - m)|.$$

Proposition 1 a) implies

$$|g_m| \le CK^m\Gamma(1 + m/k) \quad (m \ge 0),$$

hence

$$|r_{fg}(z, N)| \le C^2 K^N \sum_{m=0}^{N} \Gamma(1 + m/k)\Gamma(1 + (N - m)/k).$$

From the hint to Exercise 6 in Section 1.3 we obtain, with $\tilde{C}, \tilde{K} > 0$ sufficiently large,

$$|r_{fg}(z, N)| \le \tilde{C}\tilde{K}^N\Gamma(1 + N/k)$$

for every $N$ and $z \in \overline{S_1}$. $\qquad\square$

**Theorem 2.** *Given $k > 0$ and a sector $S$, suppose that*

$$f(z) \cong_k \hat{f}(z) \quad \text{in} \quad S.$$

*Then*

$$f'(z) \quad \cong_k \quad \hat{f}'(z) \quad \text{in} \quad S,$$

$$\int_0^z f(w)dw \quad \cong_k \quad \int_0^z \hat{f}(w)dw,$$

*with $\int_0^z \hat{f}(w)dw$ denoting the termwise integrated formal series.*

**Proof.** Given any closed subsector $\overline{S_1}$ of $S$, there exists $\delta > 0$ such that for every $z \in \overline{S_1}$, the discs around $z$ with radius $\delta|z|$ all lie in another (larger) closed subsector $\overline{S_2}$ of $S$. By assumption there exist $C, K > 0$ such that for $N \geq 0$ and $z \in \overline{S_2}$

$$|r_f(z, N)| \leq CK^N \Gamma(1 + N/k) .$$

The identity

$$r_{f'}(z, N) \quad = \quad z\frac{d}{dz}r_f(z, N + 1) + (N + 1)r_f(z, N + 1)$$

$$= \quad \frac{z}{2\pi i} \oint_{|w-z|=\delta|z|} \frac{r_f(w, N + 1)}{(w - z)^2}dw + (N + 1)r_f(z, N + 1) ,$$

valid for $N \geq 0$ and $z \in \overline{S_1}$, then implies

$$|r_{f'}(z, N)| \quad \leq \quad (\frac{1}{\delta} + N + 1)CK^{N+1}\Gamma(1 + (N + 1)/k)$$

$$\leq \quad \tilde{C}\tilde{K}^N \Gamma(1 + N/k)$$

for sufficiently large $\tilde{C}, \tilde{K} > 0$, which proves $f'(z) \cong_k \hat{f}'(z)$ in $S$. The proof of the second statement is left as an exercise. □

**Exercises.**

1. Let $k > 0$, let $\hat{f} \in \mathbb{C}[[z]]$, and let $f(z)$ be analytic in a sector $S$.

   a) Assume that to every closed subsector $\overline{S_1}$ of $S$ there exist $C, K > 0$ such that *for every $z \in \overline{S_1}$ and every natural number $N$ of the form $N = pM + q$* (with given integers $p \geq 1$ and $q \geq 0$ and arbitrary natural $M$) we have

   $$|r_f(z, N)| \leq CK^N \Gamma(1 + N/k) .$$

   Show that then $f(z) \cong_k \hat{f}(z)$ in $S$.

   b) Given $\varepsilon > 0$, assume that to every closed subsector $\overline{S_1}$ of $S$ with radius of $\overline{S_1}$ smaller than $\varepsilon$, there exist $C, K > 0$ such that *for every $z \in \overline{S_1}$ and every non-negative integer $N$* we have

   $$|r_f(z, N)| \leq CK^N \Gamma(1 + N/k) .$$

   Show that then $f(z) \cong_k \hat{f}(z)$ in $S$.

2. Let $f(z)$ be analytic in $S$, let $p$ be a natural number, and let $\hat{f}(z)$ be a formal power series. For

   $$\tilde{S} = \{z|z^p \in S\} ,$$

   define $g(z) = f(z^p)$, $z \in \tilde{S}$, and $\hat{g}(z) = \hat{f}(z^p)$. Show that then $f(z) \cong_k \hat{f}(z)$ in $S$ iff $g(z) \cong_{pk} \hat{g}(z)$ in $\tilde{S}$.

   *Hint:* You may wish to use Ex. 1.

3. Suppose $f(z) \cong_k \hat{f}(z)$ in $S$, and let the constant term of $\hat{f}(z)$ be zero. Show that then $z^{-1}f(z) \cong_k z^{-1}\hat{f}(z)$ in $S$.

4. Suppose $f(z) \cong_k \hat{f}(z)$ in $S(d, \alpha, \rho)$. Show that

$$f(ze^{\pm 2\pi i}) \cong_k \hat{f}(z) \quad \text{in} \quad S(d \mp 2\pi, \alpha, \rho).$$

# 1.6  More on Gevrey Asymptotics

The following result characterizes the units in $A_k(S)$:

**Theorem 3.**  *Given $k > 0$ and a sector $S$, suppose that*

$$f(z) \cong_k \hat{f}(z) \quad in \quad S.$$

*Moreover, assume $f(z) \neq 0$ for every $z \in .S$, and let $f_0$, the constant term of $\hat{f}(z)$, be non-zero. Then*

$$1/f(z) \cong_k 1/\hat{f}(z) \quad in \quad S,$$

*if $1/\hat{f}(z)$ denotes the unique formal power series $\hat{g}(z)$ with $\hat{f}(z)\hat{g}(z) = \hat{e}$, the unit element of $\mathbb{C}[[z]]$.*

**Proof.**  From Exercise 7 in Section 1.3 we conclude that $1/\hat{f}(z) = \sum_0^\infty \tilde{f}_n z^n \in \mathbb{C}[[z]]_{1/k}$. Given a closed subsector $\overline{S_1}$ of $S$, there exist $C, K > 0$ such that for $N \geq 0$ and $z \in \overline{S_1}$

$$|r_f(z, N)| \leq CK^N\Gamma(1 + N/k),$$

and (enlarging $C, K$ if necessary) at the same time

$$|\tilde{f}_n| \leq CK^n\Gamma(1 + n/k) \quad \text{for every} \quad n \geq 0.$$

The identity

$$r_{1/f}(z, N)f(z) = -\sum_{m=0}^{N-1} \tilde{f}_m r_f(z, m - N),$$

and the fact that for $z \in \overline{S_1}$ we have

$$|f(z)| \geq \delta > 0,$$

then imply

$$|r_{1/f}(z, N)| \leq \frac{C^2}{\delta}K^N \sum_{m=0}^{N-1} \Gamma(1 + m/k)\Gamma(1 + (m - N)/k),$$

and in the same manner as in the proof of Theorem 1 we obtain for sufficiently large $\tilde{C}, \tilde{K} > 0$

$$|r_{1/f}(z, N)| \leq \tilde{C}\tilde{K}\Gamma(1 + N/k)$$

for every $z \in \overline{S_1}$ and $N \geq 0$. $\qquad\square$

**Exercises.**

1. Given a *formal Laurent series*

$$\hat{f}(z) = \sum_{n=-m}^{\infty} f_n z^n,$$

for some natural number $m$, show that the following two statements are equivalent:

a) $f(z) - \sum_{n=-m}^{-1} f_n z^n \cong_k \sum_{n=0}^{\infty} f_n z^n$ in $S$,

b) $z^m f(z) \cong_k \sum_{n=0}^{\infty} f_{n-m} z^n$ in $S$,

and let either one serve as definition for

$$f(z) \cong_k \hat{f}(z) \quad \text{in} \quad S.$$

2. Let $\boldsymbol{L}_k(S)$ denote the set of functions $f$ which are meromorphic in $S$, so that the number of poles of $f$ in an arbitrary closed subsector of $S$ is finite, and such that

$$f(z) \cong_k \hat{f}(z)$$

for some formal Laurent series $\hat{f}$. Show that $\boldsymbol{L}_k(S)$ is a *differential field* (i.e. a differential algebra in which each element is invertible).

*Hint.* Show that the number of zeros of $f$ in an arbitrary closed subsector of $S$ is finite.

# Chapter 2

# Laplace and Borel Transforms

Laplace and (as its inverse) Borel Transform are used in several different areas of mathematics, and are therefore defined and studied in many textbooks, such as [Wi] and others. They will also prove the main tools in our theory of multisummability. However, here we may restrict to apply them to functions which are analytic in sectors and continuous at the origin, which simplifies most of the proofs for their properties. On the other hand, it will be convenient to slightly adjust the definition of Laplace Transform, so that (integer) powers are mapped to the same power times a constant. For these reasons, we choose to include all the proofs concerning properties of these operators.

## 2.1 Laplace Transforms

Let $S = S(d, \alpha)$ be a sector of infinite radius, and let $f$ be analytic and of exponential size at most $k > 0$ in $S$, and continuous at the origin. For $\tau$ with $|d - \tau| < \alpha$, the integral

$$\int_0^{\infty(\tau)} f(u) \exp\{-(u/z)^k\} d(u^k)$$

(with integration along $\arg u = \tau$) converges absolutely and compactly in the open set

$$\cos(k[\tau - \arg z]) > c|z|^k \,,$$

if $c$ is taken sufficiently large, depending upon $f$ and $\tau$. On the Riemann surface of the Logarithm, the region described by this inequality generally has infinitely many connected components, one of which is specified by the inequalities

$$-\pi/2 < k(\tau - \arg z) < \pi/2, \quad \cos(k[\tau - \arg z]) > c|z|^k \,,$$

and in this region, the function

$$g(z) = z^{-k} \int_0^{\infty(\tau)} f(u) \exp\{-(u/z)^k\} d(u^k)$$

is analytic in $z$. According to Exercise 1 in 1.2, we can assure the constant $c$ to be independent of $\tau$, as long as $\tau$ is restricted to closed subintervals of

$(d - \alpha, \ d + \alpha)$. Consequently, a change of $\tau$ essentially means a rotation of the region of convergence of the above integral, i.e. an analytic continuation of $g$. Therefore, $g(z)$ *is independent of* $\tau$. For short, we write

$$g = \mathcal{L}_k f$$

and call $\mathcal{L}_k f$ the *Laplace Transform (with index $k$) of $f$*. Altogether, one can see that $\mathcal{L}_k f$ is defined and analytic in a region which *contains a sector $\tilde{S}$ of opening smaller than, but arbitrarily close to* $\alpha + \pi/k$ ($\alpha$ being the opening of the sector $S$). In general, the radius of $\tilde{S}$ will be finite, and has to be taken smaller if one wishes the opening to become larger. For our purposes it will, most of the time, suffice bearing in mind that $\mathcal{L}_k f$ is defined and analytic (at least) in a sector $\tilde{S}$ of *opening larger than $\pi/k$* (and finite radius), *and the bisecting direction of $\tilde{S}$ being $d$,* i.e. *the same as the bisecting direction of $S$.*

One immediately sees that $g(z)$ being the Laplace Transform of index $k$ of $f(u)$ is equivalent to $g(z^{1/k})$ being the Laplace Transform of index $1$ of $f(u^{1/k})$ (in appropriate sectors). The reason for the factor $z^{-k}$ in front of the integral representation of $g(z)$ lies in the fact that we want the Laplace Transform of a power of $u$ to be the same power of $z$ times a constant (see Ex. 1, below): For $f(u) = u^{\lambda}$ (with complex $\lambda$, $\mathrm{Re}\,\lambda \geq 0$), we have

$$(\mathcal{L}_k f)(z) = \Gamma(1 + \lambda/k) z^{\lambda} \ .$$

Given a formal power series $\hat{f}(u) = \sum_0^{\infty} f_n u^n$, *termwise application* of Laplace Transform with index $k > 0$, produces the formal power series $\hat{g}(z) = \sum_0^{\infty} f_n \Gamma(1 + n/k) z^n$ $=: \hat{\mathcal{L}}_k \hat{f}$, and we frequently call $\hat{\mathcal{L}}_k$ the *formal Laplace Transform* (with index $k$).

**Theorem 1.**  *Let $f$ be analytic and of exponential size at most $k$ in a sector $S = S(d, \alpha)$, and let $g = \mathcal{L}_k f$ be its Laplace Transform with index $k$. For $k_1 > 0$, assume*

$$f(z) \ \tilde{=}_{k_1} \ \hat{f}(z) \quad in \quad S \ ,$$

*take $k_2$ so that*

$$1/k_2 \ = \ 1/k + 1/k_1 \ ,$$

*and let*

$$\hat{g} \ = \ \hat{\mathcal{L}}_k \hat{f} \ .$$

*Then to every $\varepsilon > 0$ there exists $\rho = \rho(\varepsilon) > 0$ such that $g$ is analytic in $S_{\varepsilon} = S(d, \alpha + \pi/k - \varepsilon, \rho)$, and*

$$g(z) \ \tilde{=}_{k_2} \ \hat{g}(z) \quad in \quad S_{\varepsilon} \ .$$

**Proof.**  For fixed $\delta > 0$, let $\overline{S_{\delta}} = \overline{S(d, \alpha - \delta)}$ and $\overline{S_{\delta,1}} = \overline{S(d, \alpha - \delta, 1)}$. Then by assumption (compare also Ex. 1 in 1.2), there exist $c_1, c_2$ such that

$$|f(z)| \leq c_1 \exp\{c_2 |z|^k\} \quad in \quad \overline{S_{\delta}} \ ,$$

and for suitably large $C$ and $K$, and every $N \geq 0$,

$$|r_f(z, N)| \leq CK^N \Gamma(1 + N/k_1) \quad \text{in} \quad \overline{S_{\delta,1}} .$$

From the second estimate and Proposition 1 a) we conclude

$$|f_n| \leq CK^n \Gamma(1 + n/k_1) \quad \text{for} \quad n \geq 0 .$$

Using this, one can show for $z \in \overline{S_\delta}$, $|z| \geq 1$, the existence of some $\tilde{C}, \tilde{K} > 0$ so that for every $N \geq 0$

$$\exp\{-c_2 |z|^k\} |r_f(z, N)| \leq \tilde{C} \tilde{K}^N \Gamma(1 + N/k_1) ,$$

and the same estimate holds for $z \in \overline{S_\delta}$, at least if we take $\tilde{C} \geq C$, $\tilde{K} \geq K$. Observing that $z^N r_g(z, N)$ is the Laplace Transform with index $k$ of $z^N r_f(z, N)$, one can now easily show in every closed subsector of $S_\varepsilon$, with $\delta < \varepsilon$,

$$|r_g(z, N)| \leq \hat{C} \hat{K}^N \Gamma(1 + N/k) \Gamma(1 + N/k_1)$$

for sufficiently large $\hat{C}, \hat{K}$, depending upon the subsector, and every $N \geq 0$, and with help of Stirling's Formula, one can then complete the proof. $\square$

**Remark.** It will be convenient to observe that Theorem 1 and its proof remain correct for $k_1 = \infty$ (hence $k_2 = k$), if we interprete $f(z) \stackrel{\sim}{=}_\infty \hat{f}(z)$ in $S$, according to the estimate $|r_f(z, N)| \leq CK^N$, for $N \geq 1$, to mean that $\hat{f}(z)$ converges to $f(z)$, for $z \in S$ with $|z|$ sufficiently small.

**Exercises.**

1. For $f(u) = u^\lambda$, with complex $\lambda$, $\operatorname{Re} \lambda \geq 0$, prove that

$$(\mathcal{L}_k f)(z) = \Gamma(1 + \lambda/k) z^\lambda .$$

Throughout the following three exercises, let $f(u)$ be continuous for $u = re^{i\tau}$, with real numbers $\tau$ (fixed) and $r$, $0 \leq r \leq \rho < \infty$. Define

$$g(z) = z^{-k} \int_0^{\rho e^{i\tau}} f(u) \exp\{-(u/z)^k\} d(u^k)$$

(integrating along $\arg u = \tau$).

2. Show that $g(z)$ is analytic on the Riemann surface of the Logarithm.
(In the literature, $g$ is named *finite Laplace Transform of $f$ with index $k$*.)

3. Assume that complex numbers $f_n$ and real numbers $C_n \geq 0$ (for $n \geq 0$) exist so that for every $N \geq 0$ and every $u = re^{i\tau}$, $0 \leq r \leq \rho$,

$$\left| f(u) - \sum_{n=0}^{N-1} f_n u^n \right| \leq C_N |u|^N .$$

For $\varepsilon > 0$ given, let $z$ be so that

$$\cos(k[\tau - \arg z]) \geq \varepsilon .$$

Show for $N \geq 0$

$$\left| g(z) - \sum_{n=0}^{N-1} f_n \Gamma(1 + n/k) z^n \right| \leq K_N |z|^N ,$$

with

$$K_N = \varepsilon^{-1 - N/k} \Gamma(1 + N/k) \sum_{n=0}^{N} C_n \rho^{n-N} .$$

*Hint:* Show $|f_n| \leq C_n$ for all $n \geq 0$ (compare Proposition 1 a) and

$$\left| g(z) - \sum_{n=0}^{N-1} f_n \Gamma(1 + n/k) z^n \right|$$

$$\leq |z|^{-k} \int_0^\rho \left| f(re^{i\tau}) - \sum_{n=0}^{N-1} f_n (re^{i\tau})^n \right| \exp\{ -\varepsilon (r/|z|)^k \} d(r^k)$$

$$+ \sum_{n=0}^{N-1} |f_n| \, |z|^{-k} \int_\rho^\infty r^n \exp\{ -\varepsilon (r/|z|)^k \} d(r^k)$$

(for $N \geq 0$ and $z$ as above), and observe

$$|f_n| \leq C_n \left( \frac{r}{\rho} \right)^{N-n}, \qquad r \geq \rho, \ \ 0 \leq n \leq N - 1 .$$

4. With $C_n, K_n$ as in Exercise 3, assume for $k_1 > 0$

$$C_n \leq C K^n \Gamma(1 + n/k_1), \quad n \geq 0 ,$$

with sufficiently large $C, K \geq 0$ (independent of $n$). Show that then

$$K_n \leq \tilde{C} \tilde{K}^n \Gamma(1 + n/k_2), \quad n \geq 0, \quad \text{with} \quad 1/k_2 = 1/k + 1/k_1$$

and sufficiently large $\tilde{C}, \tilde{K} \geq 0$ (independent of $n$, but depending on $\varepsilon > 0$). Compare this to Theorem 1.

5. Let $f(u)$ be continuous for $u = re^{i\tau}$, $0 \leq r \leq \infty$, and assume for sufficiently large $c_1, c_2 > 0$, and $k > 0$:

$$|f(u)| \leq c_1 \exp\{c_2|u|^k\},$$

so that $(\mathcal{L}_k f)(z)$ is defined and analytic for

$$-\pi/2 < k(\tau - \arg z) < \pi/2, \quad \cos(k[\tau - \arg z]) > c_2 r^k.$$

Show that $(\mathcal{L}_k f)(z) \equiv 0$ implies $f(u) \equiv 0$.

*Hint:* Make a change of variables $u = -(\log x)e^{i\tau}$ and use Weierstrass' Approximation Theorem. The same conclusion even holds if $\mathcal{L}_k f$ vanishes for certain discrete values $z$ accumulating at the origin; the proof essentially is along the same line.

## 2.2 Gevrey Asymptotics in Sectors of Small Opening

The following is a version of Ritt's Theorem, adapted to the theory of Gevrey asymptotics:

**Proposition 1.** *Let $\hat{f}(z) \in \mathbb{C}[[z]]_{1/k}$ and a sector $S$ of opening at most $\pi/k$ be arbitrarily given. Then there exists a function $f(z)$, analytic in $S$, so that*

$$f(z) \cong_k \hat{f}(z) \quad in \quad S.$$

**Proof.** Let $\hat{f}(z) = \sum_0^\infty f_n z^n$ and define

$$g(u) = \sum_0^\infty f_n u^n / \Gamma(1 + n/k),$$

then $\hat{f} \in \mathbb{C}[[z]]_{1/k}$ implies analyticity of $g(u)$, for $|u|$ sufficiently small. Let $d$ be the bisecting direction of $S$, and define for sufficiently small $\rho > 0$

$$f(z) = z^{-k} \int_0^{\rho e^{id}} g(u) \exp\{-(u/z)^k\} d(u^k).$$

Then one easily concludes from the Exercises at the end of the previous section that $f(z)$ has the desired properties. $\qquad \square$

**Remark.**  The above result ensures that the homomorphism $J : A_k(S) \to \mathbb{C}[[z]]_{1/k}$ (compare 1.4, 1.5) is surjective, if the opening of $S$ is smaller than or equal to $\pi/k$. It is, however, in this case not injective, as one learns from the following

**Exercises.**

1. For $k > 0$ and $c > 0$, let $f(z) = \exp\{-cz^{-k}\}$. Show that $f(z) \cong_k \hat{0}$ in $S(0, \pi/k)$ (with $\hat{0}$ being the zero power series).

2. Use the previous Ex. to conclude that to every sector $S$ of opening not more than $\pi/k$ there exists $f(z)$, analytic and non-zero in $S$, with

$$f(z) \cong_k \hat{0} \quad \text{in} \quad S.$$

3. Let $S$ be a sector of opening not more than $\pi/k$, and let $f(z)$ be analytic in $S$ with

$$f(z) \cong_k \hat{0} \quad \text{in} \quad S.$$

   To each closed subsector $\overline{S_1}$ of $S$, find $c_1, c_2 > 0$ so that

$$|f(z)| \le c_1 \exp\{-c_2|z|^{-k}\} \quad \text{in} \quad \overline{S_1}.$$

   *Hint:*  For sufficiently large $C, K > 0$, show

$$|f(z)| \le C(|z|K)^N N^{N/k} \quad \text{in} \quad \overline{S_1},$$

   for every $N \ge 0$, and then (for fixed $|z|$) determine $N$ so that the right hand side is (approximately) minimal.

4. Let $S$ be a sector of opening not more than $\pi/k$, and let $f$ be analytic in $S$ with

$$f(z) \cong_k \hat{0} \quad \text{in} \quad S.$$

   Moreover, let $g(z)$ be analytic in $S$ and so that for some real $c > 0$ we have $z^c g(z)$ bounded at the origin. Show

$$f(z)g(z) \cong_k \hat{0} \quad \text{in} \quad S.$$

## 2.3  Borel Transforms

For arbitrary real $\tau$, and $k, \varepsilon > 0$, let $\gamma_k(\tau)$ denote the path from the origin along $\arg z = \tau + (\varepsilon + \pi)/(2k)$ to some finite point $z_1$, then along the circle $|z| = |z_1|$ to the ray $\arg z = \tau - (\varepsilon + \pi)/(2k)$, and back to the origin along this ray (the dependence of the path on $\varepsilon$ and the point $z_1$ will be inessential, so we do not display this dependence, to keep notation simple). Let $S = S(d, \alpha, \rho)$ be a sector of opening $\alpha > \pi/k$. Then for $\tau$ with

$$|\tau - d| < (\alpha - \pi/k)/2 \, ,$$

we may choose $\varepsilon$ and $z_1$ so that $\gamma_k(\tau)$ fits into the sector $S$. If $f(z)$ is analytic in $S$ and bounded at the origin, we define (with $\varepsilon \le \pi$) the Borel Transform of $f$ with index $k$ by

$$(\mathcal{B}_k f)(u) = \frac{1}{2\pi i} \int_{\gamma_k(\tau)} z^k f(z) \exp\{(u/z)^k\} d(z^{-k}) \, ,$$

for $u \in S(\tau, \varepsilon/k)$ (observe that then the exponential function in the integral decreases along the two radial parts of the path, so that the integral converges absolutely and compactly and represents an analytic function of $u$). From Cauchy's Theorem we conclude that $\mathcal{B}_k f$ is independent of $\varepsilon$ and $z_1$, and a change of $\tau$ gives the analytic continuation of $\mathcal{B}_k f$. So we find that $\mathcal{B}_k f$ is also independent of $\tau$, and analytic in the sector

$$S(d, \alpha - \pi/k) \, .$$

The exercise (below) shows that termwise application of $\mathcal{B}_k$ to a formal power series $\hat{f}(z) = \sum_0^\infty f_n z^n$ produces the formal power series

$$(\hat{\mathcal{B}}_k \hat{f})(z) = \sum_0^\infty f_n z^n / \Gamma(1 + n/k) \, ,$$

called the formal Borel Transform of $\hat{f}$.

**Theorem 2.**  Let $S = S(d, \alpha, \rho)$ be an arbitrary sector, let $f$ be analytic in $S$, and (for $k_1 > 0$)

$$f(z) \cong_{k_1} \hat{f}(z) \quad in \quad S \, .$$

Let $k > 0$ be such that

$$\alpha > \pi/k \, ,$$

so that $\mathcal{B}_k f$ is defined and analytic in $\tilde{S} = S(d, \alpha - \pi/k)$. Define $k_2$ by

$$k_2^{-1} = k_1^{-1} - k^{-1}, \quad if \quad k_1 < k \, ,$$

resp.

$$k_2 = \infty \quad if \quad k_1 \ge k \, .$$

Then

$$(\mathcal{B}_k f)(u) \cong_{k_2} (\hat{\mathcal{B}}_k \hat{f})(u) \quad in \quad \tilde{S} \, .$$

**Proof.** For $\tau$ sufficiently close to $d$ and $\gamma_k(\tau)$ in $S$, we have for sufficiently large $C, K > 0$ :

$$|r_f(z, N)| \leq CK^N \Gamma(1 + N/k_1) \,,$$

for every $N \geq 0$ and every $z \in \gamma_k(\tau)$. With $g = \mathcal{B}_k f$ we find that

$$u^N r_g(u, N) = \left(\mathcal{B}_k z^N r_f(z, N)\right)(u) \,, \quad u \in S(\tau, \varepsilon/k) \,.$$

Breaking $\gamma_k(\tau)$ into three pieces (the two radial parts and the circular arc of radius $|z_1| = r$ and estimating the resulting three integrals in the usual way, we find for $N \geq 0$ and $u \in S(\tau, \varepsilon/(2k))$

$$|u^N r_g(u, N)| \leq CK^N (2\pi)^{-1} \Gamma(1 + N/k_1)(I_1 + I_2 + I_3) \,,$$

with

$$I_1, I_3 \leq k \int_0^r x^{N-1} \exp\{-c(|u|/x)^k\} dx \,, \quad I_2 \leq \tilde{c} r^N \exp\{(|u|/r)^k\} \,,$$

for suitable constants $c, \tilde{c} > 0$ (independent of $u, N$, and $r$). It suffices to consider large $N$; the final estimate then holds automatically for all $N$, with possibly enlarged constants. For large $N$, we may choose $\gamma_k(\tau)$ so that, for fixed $u$, $r = |u| (N/k)^{-1/k}$. This gives

$$I_2 \leq \tilde{c} |u|^N (N/k)^{-N/k} e^{N/k}$$

and (after substituting $y = c(|u|/x)^k$)

$$I_1, I_3 \leq |u|^N c^{N/k} \int_{cN/k}^\infty y^{-1-N/k} e^{-y} dy \leq |u|^N c^{-1}(N/k)^{-1-N/k} \int_0^\infty e^{-y} dy \,.$$

Using Stirling's Formula, this shows, with suitably large $\tilde{C}, \tilde{K} > 0$, independent of $u$ and $N$,

$$|r_g(u, N)| \leq \tilde{C} \tilde{K}^N \Gamma(1 + N/k_1)/\Gamma(1 + N/k) \,.$$

It is then easy to complete the proof.                                    □

**Remark.** In case $k_1 \geq k$, Theorem 2 says that $\mathcal{B}_k f$ is analytic at the origin, and $\hat{\mathcal{B}}_k \hat{f}$ is its power series expansion. If $k_1 > k$, one can see (either from the final estimate in the proof or by looking at $\hat{\mathcal{B}}_k \hat{f}$ and estimating its coefficients) that $\mathcal{B}_k f$ is an entire function of exponential size not larger than $(k^{-1} - k_1^{-1})^{-1}$.

**Exercise.**

For $f(z) = z^\lambda$ ($\lambda \in \mathbb{C}$), use Hankels integral representation of the reciprocal Gamma function to show $(\mathcal{B}_k f)(u)$ exists and equals $u^\lambda/\Gamma(1 + \lambda/k)$, for every $k > 0$.

## 2.4  Inversion Formulas

The following two Theorems say that, roughly speaking, $\mathcal{B}_k$ is the inverse operator of $\mathcal{L}_k$, on suitable function spaces. A closer look shows that, in general, $\mathcal{L}_k(\mathcal{B}_k f)(z)$ will be defined on a somewhat smaller set of $z$ than is the original $f(z)$. However, since we are dealing with analytic functions, this is inessential.

**Theorem 3.**  *Under the assumptions of Theorem 2, let*

$$g(u) = (\mathcal{B}_k f)(u), \quad u \in \tilde{S}.$$

*Then $g(u)$ is of exponential size not more than $k$ in $\tilde{S}$, so that $(\mathcal{L}_k g)(z)$ is analytic in a sector $\hat{S} = S(d, \hat{\alpha}, \hat{\rho})$ (with $\pi/k < \hat{\alpha} \leq \alpha$, $\hat{\rho} \leq \rho$), and*

$$f(z) = (\mathcal{L}_k g)(z), \quad z \in \hat{S}.$$

**Proof.**  If we split $\gamma_k(\tau)$ into three pieces, as in the proof of Theorem 2, then the integrals over the radial parts uniformly tend to zero as $z \to \infty$ in $S(\tau, \varepsilon/(2k))$. In the integral over the circular arc, one may expand $\exp\{(u/z)^k\}$ into the exponential series and integrate termwise to see that this integral is an entire function of $u$ of exponential size not more than $k$. Altogether, this shows $\mathcal{B}_k f$ of exponential size not more than $k$, in $S(\tau, \varepsilon/(2k))$. Varying $\tau$, we find the same in $\tilde{S}$.

To prove $f = \mathcal{L}_k g$, it suffices to do so for $z$ with $\arg z = d$ and $|z|$ sufficiently small. For such $z$, one can in $\mathcal{L}_k g = \mathcal{L}_k(\mathcal{B}_k f)$ (with $\tau = d$) interchange the order of integration, giving

$$
\mathcal{L}_k(\mathcal{B}_k f)(z) = z^{-k} \frac{1}{2\pi i} \int_{\gamma_k(d)} w^k f(w) \int_0^{\infty(d)} \exp\{u^k(w^{-k} - z^{-k})\} d(u^k) d(w^{-k})
$$

$$
= \frac{-k}{2\pi i} \int_{\gamma_k(d)} \frac{w^{k-1} f(w)}{w^k - z^k} dw.
$$

The function $F_z(w) = w^{k-1} f(w)(w^k - z^k)^{-1}$ can be seen to have, in the interior of $\gamma_k(d)$, exactly one singularity, namely at $w = z$, this being a pole of order one with residue $f(z)/k$. Since $\gamma_k(d)$ has negative orientation, the Residue Theorem completes the proof. □

**Theorem 4.**  *Under the assumptions of Theorem 1, the function $(\mathcal{B}_k g)(u)$ is analytic in $S$, and*

$$f(u) = (\mathcal{B}_k g)(u), \quad u \in S.$$

**Proof.**  From Theorems 1 and 2 we obtain that $\tilde{f}(u) := (\mathcal{B}_k g)(u)$ is analytic in $S$, and Theorem 3 implies that $(\mathcal{L}_k \tilde{f})(z) = g(z) = (\mathcal{L}_k f)(z)$, for $z$ in some sector bisected by $d$. From Ex. 5, Section 2.1, we learn that this implies $\tilde{f}(u) = f(u)$, $u \in S$. □

**Exercises.**

1. Show that in Theorems 3 and 4 it suffices to prove the case $k = 1$; the general case then follows through suitable changes of variables.

2. Give a different proof for Theorem 4, in case of $k = 1$, by suitably deforming the path of integration in the Borel transform, and then interchanging the order of integration.

# Chapter 3

# Summable Power Series

The concept of $k$-summability of formal power series had been "in the air", long before J.P. Ramis in the '70s gave the formal definition:

Between 1910 and 1920, G.N. Watson and F. Nevanlinna discovered that Gevrey asymptotics in sectors of large opening characterize the corresponding analytic function uniquely (compare Theorem 1 below), and that this function can be represented as a Laplace integral over another function which is analytic at the origin. J. Horn used Laplace Transform to represent solutions to certain linear systems of ODE's, but did not relate his results to the ones of Watson and Nevanlinna. Trjitzinsky (1935) and Turrittin (1955) applied the same techniques to slightly more general systems and pointed out the limitation of this approach to special situations. In a series of papers [Ba 3–6], the author defined and studied *first level formal solutions*, and only later realized that these were always summable in Ramis' sense.

## 3.1 Gevrey Asymptotics in Sectors of Large Opening

The following result essentially is equivalent to the *injectivity* of $J : A_k(S) \rightarrow \mathbb{C}[[z]]_{1/k}$ for sectors $S$ *of opening more than* $\pi/k$.

**Theorem 1.** *Let* $k > 0$ *and* $\hat{f} \in \mathbb{C}[[z]]_{1/k}$. *Then* $\hat{g}(u) = (\hat{B}_k \hat{f})(u)$ *converges, for* $|u|$ *sufficiently small, and we define* $g = S\hat{g}$ *(so that* $g(u)$ *is analytic in some neighborhood of the origin). Then the following two statements are equivalent, for every fixed real* $d$ :

a) *There exists a sector* $S = S(d, \alpha, \rho)$ *with* $\alpha > \pi/k$, *and* $f \in A_k(S)$ *with* $\hat{f} = J(f)$.

b) *There exists a sector* $\tilde{S} = S(d, \varepsilon)$ *so that* $g$ *admits analytic continuation into* $\tilde{S}$ *and is of exponential size not more than* $k$ *there.*

*Moreover, if a) or b) holds, then* $f = \mathcal{L}_k g$ *follows, so that* $f$ *is uniquely determined by* $\hat{f}$.

**Proof.**   Follows immediately from Theorems 1, 2, 3 in Chapter 2.                    □

**Remark.**   The following exercise gives an example of an $\hat{f} \in \mathbb{C}[[z]]_{1/k}$ so that the sum of $\hat{g} = \hat{\mathcal{B}}_k \hat{f}$ cannot be continued beyond its circle of convergence. This shows that for sectors of opening more than $\pi/k$, the map $J : A_k(S) \rightarrow \mathbb{C}[[z]]_{1/k}$ will not be surjective.

**Exercise.**   For $\hat{f}(z) = \sum_0^\infty \Gamma(1 + 2^n/k) z^{2^n}$, $k > 0$, show that $\hat{g} = \hat{\mathcal{B}}_k \hat{f}$ has radius of convergence equal to one. Moreover, show for the sum $g = \mathcal{S}\hat{g}$ the functional equations

$$g(z) = \sum_{\nu=0}^{m-1} z^{2^\nu} + g(z^{2^m}), \quad m \geq 1,$$

and use these to show that $g(z)$ cannot have a radial limit at points $z_{jm} = \exp\{2\pi i j 2^{-m}\}$. From this fact, conclude that $g$ cannot be analytically continued beyond the unit disc.

## 3.2   Definition of k-Summability in a Direction  d

Let $k > 0$, $d \in \mathbb{R}$, and $\hat{f} \in \mathbb{C}[[z]]$ be given. We say that $\hat{f}$ is $k$-*summable in direction* $d$, if a sector $S = S(d, \alpha, \rho)$, with $\alpha > \pi/k$, and a function $f \in A_k(S)$ exist with $J(f) = \hat{f}$. If this is so, then $\hat{f} \in \mathbb{C}[[z]]_{1/k}$ follows (see Section 1.5), and Theorem 1 guarantees uniqueness of $f \in A_k(S)$. In fact, the analyticity of $f$ even shows that $f$ does not really depend on $S$ (but may depend on the direction $d$, as can be seen from examples given later). In view of Theorem 1, $\hat{f}$ is $k$-summable in direction $d$ iff $\hat{g} = \hat{\mathcal{B}}_k \hat{f}$ converges and $g = \mathcal{S}(\hat{g})$ is analytic and of exponential size at most $k$ in a sector $\tilde{S} = S(d, \varepsilon)$ (for some $\varepsilon > 0$), and then $f = \mathcal{L}_k g$. If $\hat{f}$ is $k$-summable in direction $d$, we call the unique function $f = \mathcal{L}_k g$ the $k$-*sum of $\hat{f}$ in direction* $d$, and we write

$$f = \mathcal{S}_{k,d}(\hat{f}) .$$

**Lemma 1.**

a) *Let $\hat{f}$ be convergent. Then for every $k > 0$ and every $d$, $\hat{f}$ is $k$-summable in direction $d$, and*

$$\mathcal{S}_{k,d}(\hat{f})(z) = \mathcal{S}(\hat{f})(z)$$

   *for every $z$, on the Riemann surface of the Logarithm, where both sides are defined .*

b) *Let $\hat{f}$ be $k$-summable in direction $d$, and let $\varepsilon > 0$ be sufficiently small. Then $\hat{f}$ is $k$-summable in all directions $\tilde{d}$ with*

$$|\tilde{d} - d| < \varepsilon .$$

c) Let $\hat{f}$ be $k$-summable in direction $d$, and let $\varepsilon > 0$ be sufficiently small. Then $\hat{f}$ is $(k - \varepsilon)$-summable in direction $d$.

**Proof.** For a), use $f(z) = S(\hat{f})(z) \cong_k \hat{f}(z)$ in $S$, for every $k > 0$ and every sector $S$ of sufficiently small radius. To prove b), use the definition and observe

$$\alpha - \varepsilon > \pi/k, \quad S(\tilde{d}, \alpha - \varepsilon, \rho) \subset S(d, \alpha, \rho)$$

(for small $\varepsilon > 0$). For c), use

$$\alpha > \pi/(k - \varepsilon), \quad \Gamma(1 + N/k)/\Gamma(1 + N/(k - \varepsilon)) \to 0 \quad (N \to \infty)$$

(for small $\varepsilon > 0$). $\qquad\qquad\qquad\qquad\qquad\qquad\qquad\qquad\qquad\qquad\qquad\square$

**Remark.** If we are given a series $\hat{f}$ whose $k$-summability in direction $d$ is to be shown, and if we then want to compute its sum, we are presented with the following problems:

a) The function $g(z)$, locally given by the convergent series $\hat{B}_k \hat{f}$, has to be analytically continued into a (small) sector $S(d, \varepsilon)$.

b) In this sector, we have to show that $g$ is of exponential size not larger than $k$.

c) We have to compute the integral $\mathcal{L}_k g$.

At first glance, one might think that a) might be the major problem, but on one hand, there are explicit methods for performing analytic continuation (once we know that $g$ is analytic in $S(d, \varepsilon)$, and in view of Lemma 1 c), we may even replace $k$ by $\tilde{k} < k$ (with $k - \tilde{k}$ sufficiently small), since then $\hat{B}_{\tilde{k}} \hat{f}$ will converge to an entire function, so that a) is no problem at all. This entire function, however, will have too large an exponential size in, generally, all directions but $\arg u = d$. So in a way, the main difficulty lies in the problem of verifying b).

**Lemma 2.** Let $\hat{f}$ be $k_1$-summable in direction $d$, and let $k > k_1$ and $k_2$ be so that

$$1/k_2 = 1/k_1 - 1/k.$$

Then $\hat{g} = \hat{B}_k \hat{f}$ is $k_2$-summable in direction $d$, and

$$S_{k_2,d}(\hat{g}) = B_k \left( S_{k_1,d}(\hat{f}) \right).$$

**Proof.** Follows immediately from Theorem 2, Section 2.3. $\qquad\qquad\qquad\qquad\square$

**Lemma 3.**

a) *Let $\hat{f}$ be $k$-summable in direction $d$, for every $d \in (\alpha, \beta)$, $\alpha < \beta$, and fixed $k > 0$. Then for every $d_1, d_2 \in (\alpha, \beta)$*

$$\mathcal{S}_{k,d_1}(\hat{f})(z) = \mathcal{S}_{k,d_2}(\hat{f})(z) ,$$

*for every $z$ where both sides are defined.*

b) *Let $\hat{f}$ be $k_j$-summable in direction $d$, for $j = 1, 2$, with $k_1 > k_2 > 0$ and some fixed $d$, then*

$$\mathcal{S}_{k_1,d}(\hat{f})(z) = \mathcal{S}_{k_2,d}(\hat{f})(z) ,$$

*for every $z$ where both sides are defined.*

c) *For $\tilde{d} = d + 2\pi$, $k$-summability of $\hat{f}$ in direction $d$ is equivalent to $k$-summability of $\hat{f}$ in direction $\tilde{d}$, and*

$$\mathcal{S}_{k,\tilde{d}}(\hat{f})(z) = \mathcal{S}_{k,d}(\hat{f})(ze^{-2\pi i}) ,$$

*for every $z$ where both sides are defined.*

**Proof.**

a) The function $g = \mathcal{S}(\hat{B}_k \hat{f})$ is analytic and of exponential size not more than $k$, in the sector $S((\alpha + \beta)/2, \beta - \alpha)$. Hence $f = \mathcal{L}_k g \cong_k \hat{f}$ in a region which, for every $d \in (\alpha, \beta)$, contains a sector of opening larger than $\pi/k$ and bisecting direction $d$. So $\mathcal{S}_{k,d}(\hat{f}) = f$, for every such $d$.

b) Lemma 2 and $k_2$-summability in direction $d$ imply that $\hat{g} = \hat{B}_{k_1} \dot{\hat{f}}$ is $\tilde{k}$-summable in direction $d$, with $1/\tilde{k} = 1/k_2 - 1/k_1$, and

$$\mathcal{S}_{\tilde{k},d}(\hat{g}) = \mathcal{B}_{k_1}\left(\mathcal{S}_{k_2,d}(\hat{f})\right) ,$$

or equivalently,

$$\mathcal{S}_{k_2,d}(\hat{f}) = \mathcal{L}_{k_1}\left(\mathcal{S}_{\tilde{k},d}(\hat{g})\right) .$$

Since $\hat{f}$ also is $k_1$-summable in direction $d$, we have $\hat{g}$ convergent, hence from Lemma 1 a)

$$\mathcal{S}_{\tilde{k},d}(\hat{g}) = \mathcal{S}(\hat{g}) ,$$

and by definition of $k_1$-summability

$$\mathcal{S}_{k_1,d}(\hat{f}) = \mathcal{L}_{k_1}\left(\mathcal{S}(\hat{g})\right) .$$

c) Use Ex. 4, Section 1.5.                                                                $\square$

**Exercises.**

1. Show that $\hat{f}(z) = \sum_{0}^{\infty} \Gamma(1 + n/k)z^n$ is $k$-summable in every direction $d \neq 2j\pi$, $j \in \mathbb{Z}$.

   *Hint.* Compute $g = S(\hat{B}_k \hat{f})$.

2. Assume that $\hat{f}$ is so that $g = S(\hat{B}_k \hat{f})$ is a rational function. Let $d$ be so that no poles of $g$ lie on the ray $\arg u = d$, and show that then $\hat{f}$ is $k$-summable in direction $d$.

3. Show that $\hat{f}(z) = \sum_{0}^{\infty} \Gamma(1 + n/k)z^n/\Gamma(1 + n)$ is $k$-summable in directions $d$ with

$$(2j + 1/2)\pi < d < (2j + 3/2)\pi, \quad j \in \mathbb{Z} .$$

4. For $\hat{f}$ as in Ex. 3, let $d$ be such that

$$(2j - 1/2)\pi < d < (2j + 1/2)\pi, \quad j \in \mathbb{Z} .$$

   Show that $\hat{f}$ is (resp. is not) $k$-summable in direction $d$, if $k \geq 1$ (resp. $0 < k < 1$).

5. Assume $\hat{f}$ so that

$$g = S(\hat{B}_k \hat{f}) = \int_{0}^{\infty} e^{zt} t^{-t} dt .$$

   Show that $\hat{f}$ is $k$-summable in all directions $d \neq 2j\pi$, $j \in \mathbb{Z}$, and is not $k$-summable in the remaining directions.

   *Hint.* Use Ex. 2–5, Section 1.2.

6. Show that $\hat{f}(z) = \sum_{0}^{\infty} \Gamma(1 + 2n)z^n/\Gamma(1 + n)$

   a) is $1$-summable in every direction $d \in [-\pi, \pi)$ but one (which?).
   b) is $1/2$-summable in every direction $d$ with

$$(2j + 1/2)\pi < d < (2j + 3/2)\pi, \quad j \in \mathbb{Z} .$$

   *Hint.* For a), show

$$(1 - 4z)^{-1/2} = \sum_{n=0}^{\infty} \frac{\Gamma(1 + 2n)}{\{\Gamma(1 + n)\}^2} z^n, \qquad |z| < 1/4 .$$

## 3.3   The Algebra of k-Summable Series in a Direction

Let $k > 0$ and $d$ be given. By

$$\mathbb{C}\{z\}_{k,d}$$

we denote the set of all $\hat{f}$ that are $k$-summable in direction $d$. This set is a differential algebra over $\mathbb{C}$, as follows from

**Theorem 2.**  *For fixed, but arbitrary, $k > 0$ and $d$, we have:*

a) *If $\hat{f}, \hat{g} \in \mathbb{C}\{z\}_{k,d}$, then $\hat{f} + \hat{g}$, $\hat{f}\hat{g} \in \mathbb{C}\{z\}_{k,d}$, and*

$$\mathcal{S}_{k,d}(\hat{f} + \hat{g}) \quad = \quad \mathcal{S}_{k,d}(\hat{f}) + \mathcal{S}_{k,d}(\hat{g}) \, ,$$

$$\mathcal{S}_{k,d}(\hat{f}\hat{g}) \quad = \quad (\mathcal{S}_{k,d}(\hat{f}))(\mathcal{S}_{k,d}(\hat{g})) \, .$$

b) *If $\hat{f} \in \mathbb{C}\{z\}_{k,d}$, then $\hat{f}'$, $\int_0^z \hat{f}(w)dw \in \mathbb{C}\{z\}_{k,d}$, and*

$$\mathcal{S}_{k,d}(\hat{f}') \quad = \quad \frac{d}{dz}\mathcal{S}_{k,d}(\hat{f}) \, ,$$

$$\mathcal{S}_{k,d}\Big( \int_0^z \hat{f}(w)dw \Big) \quad = \quad \int_0^z \mathcal{S}_{k,d}(\hat{f})(w)dw \, .$$

c) *If $\hat{f} \in \mathbb{C}\{z\}_{k,d}$ has non-zero constant term, then $1/\hat{f} \in \mathbb{C}\{z\}_{k,d}$, and*

$$\mathcal{S}_{k,d}(1/\hat{f}) = 1/\big(\mathcal{S}_{k,d}(\hat{f})\big)$$

*(wherever the right hand side is defined).*

d) *If $\hat{f} \in \mathbb{C}\{z\}_{k,d}$ and $p$ is a natural number, then $\hat{f}(z^p) \in \dot{\mathbb{C}}\{z\}_{pk,d/p}$, and*

$$\mathcal{S}_{pk,d/p}\big(\hat{f}(z^p)\big) = \mathcal{S}_{k,d}(\hat{f})(z^p) \, .$$

**Proof.**   Follows from Theorems 1, 2, 3 in Chapter 1 and Ex. 2 in Section 1.6.

**Exercises.**

1. Let $\hat{f} \in \mathbb{C}\{z\}_{k,d}$ have zero constant term. Show that $z^{-1}\hat{f}(z) \in \mathbb{C}\{z\}_{k,d}$, and $\mathcal{S}_{k,d}\big(z^{-1}\hat{f}(z)\big) = z^{-1}\mathcal{S}_{k,d}(\hat{f})(z)$.

   *Hint.* Compare Ex. 3, Section 1.5.

2. Let $\hat{f} \in \mathbb{C}\{z\}_{k,d}$ *formally* satisfy the $n$-th order linear differential equation

$$\sum_{j=0}^{n} a_j(z) y^{(j)} = 0 \, ,$$

with coefficients $a_j(z)$ analytic near the origin. Show that $f = S_{k,d}(\hat{f})$ then is a solution of the same equation.

## 3.4   Definition of k-Summability

In view of Lemma 3 c), we will from now on *identify directions $d$ which differ by integer multiples of $2\pi$* (despite of the fact that variables $z, u$ still are considered on the Riemann surface of the Logarithm). From Lemma 1 b) we see that the set of directions $d$ for which some $\hat{f}$ is $k$-summable is always open. Examples show that the complement of this set can be uncountable; however in most applications one encounters series $\hat{f}$ *being $k$-summable in all directions $d$ but* (after identification modulo $2\pi$) *finitely many directions* $d_1, \ldots, d_n$. Whenever this is so, we simply call $\hat{f}$ *$k$-summable*. The directions $d_1, \ldots, d_n$ then are called *the singular directions of $\hat{f}$* (if they occur). For the set of all $k$-summable series $\hat{f}$ we write

$$\mathbb{C}\{z\}_k \, ,$$

and it may sometimes be convenient to define $\mathbb{C}\{z\}_\infty = \mathbb{C}\{z\}$ (the set of all convergent series).

It is immediately clear that Theorem 2 remains correct if we replace $\mathbb{C}\{z\}_{k,d}$ by $\mathbb{C}\{z\}_k$, and the identities for the respective sums then hold for every but finitely many $d$ (modulo $2\pi$).

Lemma 3 a) says that $S_{k,d}(\hat{f})$ is independent of $d$, as $d$ varies in an interval not containing singular directions. At a singular direction, however, the sum will change abruptly, as can be seen from

**Proposition 1.** *For $\alpha < d_0 < \beta$, $k > 0$, assume that $\hat{f}$ is $k$-summable in all directions $d \in (\alpha, \beta)$, $d \neq d_0$. For some $d_1, d_2$ with $\alpha < d_1 < d_0 < d_2 < \beta$, $|d_1 - d_2| < \pi/(2k)$, assume*

$$S_{k,d_1}(\hat{f})(z) = S_{k,d_2}(\hat{f})(z) \, ,$$

*for all $z$ where both sides are defined. Then $\hat{f}$ is $k$-summable in direction $d_0$.*

**Proof.** For $j = 1, 2$, the functions $f_j = S_{k,d_j}(\hat{f})$ are analytic in sectors $S_j = S(d_j, \alpha_j, \rho)$, with $\alpha_j > \pi/k$. Due to our assumptions, these sectors overlap (on the Riemann surface of the Logarithm), and within their intersection the functions are equal. Hence $f = f_1 = f_2$ is analytic in $S_1 \cup S_2$, and $f(z) \cong_k \hat{f}(z)$ in $S_1 \cup S_2$. Obviously, $S_1 \cup S_2$ contains a subsector of opening larger than $\pi/k$ and bisecting direction $d_0$, so $\hat{f}$ is $k$-summable in direction $d_0$.                    □

**Proposition 2.**   *Assume* $\hat{f} \in \mathbb{C}\{z\}_k$ *has no singular direction (in other words,* $\hat{f} \in \mathbb{C}\{z\}_{k,d}$ *for every* $d$). *Then* $\hat{f} \in \mathbb{C}\{z\}$ *(i.e.* $\hat{f}(z)$ *converges for sufficiently small* $|z|$).

**Proof.**   According to Lemma 3 a), c), the function $f = S_{k,d}(\hat{f})$ is independent of $d$ and single-valued (hence $f(ze^{2\pi i}) = f(z)$ for every $z$). From Proposition 1 b) in Chapter 1 we then obtain $\hat{f} \in \mathbb{C}\{z\}$.                                           □

Suppose $\hat{f} \in \mathbb{C}\{z\}_k$ is given, and $d_0$ is a singular direction of $\hat{f}$. Then for $g = S(\hat{\mathcal{B}}_k \hat{f}_k)$, two things can happen: either $g(u)$ is singular for some $u_0$ with $\arg u_0 = d_0$ (so that analytic continuation along $\arg u = d_0$ breaks down at this point), or otherwise (since directions $d$ close to $d_0$ are not singular) $g(u)$ is analytic in a sector $S(d_0, \alpha, +\infty)$, for some $\alpha > 0$, but is not of exponential size at most $k$ along $\arg u = d_0$. This can occur, according to Ex. 5, Section 3.2, but only if $g(u)$ is of *infinite exponential* size along $\arg u = d_0$; this can be seen from

**Proposition 3.**   *For* $k > 0$ *and* $\alpha < d_0 < \beta$, *assume that* $\hat{f}$ *is* $k$-*summable in direction* $d$, *for every* $d \neq d_0$, $d \in (\alpha, \beta)$. *Moreover, for* $\tilde{k} > k$, *assume that* $g = S(\hat{\mathcal{B}}_k(\hat{f}))$ *can be analytically continued along* $\arg u = d_0$ *and is of exponential size at most* $\tilde{k}$ *along* $\arg u = d_0$. *Then* $\hat{f}$ *is* $k$-*summable in direction* $d_0$.

**Proof.**   In $S = S((\alpha + \beta)/2, \beta - \alpha)$, we have that $g = S(\hat{\mathcal{B}}_k \hat{f})$, along every ray $\arg u = d$ but for $d = d_0$, is of exponential size at most $k$. An application of Phragmen-Lindelöf's Theorem then shows that the same holds for $d = d_0$.   □

From Ex. 6, Section 3.2, we learn that for fixed $d$, a (divergent) series $\hat{f}$ may very well be $k_1$-, and likewise $k_2$-summable in direction $d$. This changes drastically, if we require the same for all but finitely many directions at the same time:

**Theorem 3.**   *For* $k_1 > k_2 > 0$ *we have*

$$\mathbb{C}\{z\} = \mathbb{C}\{z\}_{k_2} \cap \mathbb{C}\{z\}_{k_1} = \mathbb{C}\{z\}_{k_2} \cap \mathbb{C}[[z]]_{1/k_1} \,.$$

**Proof.**   Trivially

$$\mathbb{C}\{z\} \subset \mathbb{C}\{z\}_{k_2} \cap \mathbb{C}\{z\}_{k_1} \subset \mathbb{C}\{z\}_{k_2} \cap \mathbb{C}[[z]]_{1/k_1} \,.$$

If $\hat{f} \in \mathbb{C}\{z\}_{k_2} \cap \mathbb{C}[[z]]_{1/k_1}$, then $\hat{g} = \hat{B}_{k_2}(\hat{f})$ is entire and of exponential size at most $k$ everywhere, with $1/k = 1/k_2 - 1/k_1$ (compare the Example in Section 1.2). Therefore, Proposition 3 implies that $\hat{f}$ cannot have any singular directions, hence converges because of Proposition 2. □

**Exercises.**

1. Investigate $k$-summability of the series $\hat{f}$ in the Exercises 1, 2, 3, 5, 6, Section 3.2.

2. For $k_1 > k_2 > 0$, let $\hat{f} \in \mathbb{C}\{z\}_{k_2,d} \cap \mathbb{C}[[z]]_{1/k_1}$, for every $d \in (\alpha, \beta)$, $\alpha < \beta$. Prove $\hat{f} \in \mathbb{C}\{z\}_{k_1,d}$ for every $d \in (\alpha - \pi/(2k), \beta + \pi/(2k))$, with $1/k = 1/k_2 - 1/k_1$.

3. For $\mathrm{Re}\,\lambda > 0$, show that $\hat{f}(z; \lambda) = \sum_0^\infty \Gamma(\lambda + n) z^n$ is 1-summable, its 1-sum being the function

$$f(z; \lambda) = z^{-\lambda} \int_0^{\infty(d)} \frac{w^{\lambda-1}}{1-w} e^{-w/z} dw$$

(for $d \neq 2k\pi$, $k \in \mathbb{Z}$).

4. For (complex) $\lambda \neq 0, -1, -2, \ldots$, write $\hat{f}(z; \lambda)$ in the form

$$\hat{f}(z; \lambda) = \sum_0^{m-1} \Gamma(\lambda + n) z^n + z^m \sum_0^\infty \Gamma(\lambda + m + n) z^n,$$

for sufficiently large integer $m$, and use the previous exercise to conclude $\hat{f}(z; \lambda) \in \mathbb{C}\{z\}_1$.

In the following exercises, we use the convention

$$(\alpha)_0 = 1, \quad (\alpha)_n = \alpha(\alpha + 1) \ldots (\alpha + n - 1), \quad n \geq 1.$$

Observe that $(\alpha)_n = \Gamma(\alpha + n)/\Gamma(\alpha)$, whenever that quotient gives sense, and $(1)_n = n!$

5. For $\hat{f}(z) = \sum f_n z^n \in \mathbb{C}\{z\}_1$ and complex $\alpha, \beta$ with $\beta \neq 0, -1, -2, \ldots$, show

$$\hat{f}(z; \alpha, \beta) = \sum f_n \frac{(\alpha)_n}{(\beta)_n} z^n \in \mathbb{C}\{z\}_1.$$

*Hint.* For $\mathrm{Re}\,\beta > \mathrm{Re}\,\alpha > 0$, use the Beta-Integral to show an integral formula expressing $z^{\beta-1} \hat{B}_1 \hat{f}(z; \alpha, \beta)$ (formally) in terms of $z^{\alpha-1} \hat{B}_1 \hat{f}(z)$, and then conclude $\hat{f}(z; \alpha, \beta) \in \mathbb{C}\{z\}_1$. To remove this restrictions on $\alpha$ and $\beta$, use termwise integration of $z^{\beta-1}\hat{f}(z; \alpha, \beta)$ resp. termwise differentiation of $z^{\alpha}\hat{f}(z; \alpha, \beta)$.

6. From Ex. 5, conclude that the generalized hypergeometric series

$$_pF_q(\alpha_1, \ldots, \alpha_p, \beta_1, \ldots, \beta_q; z) = \sum_{n=0}^{\infty} \frac{(\alpha_1)_n \ldots (\alpha_p)_n}{(\beta_1)_n \ldots (\beta_q)_n} \frac{z^n}{n!}$$

(with suitable restrictions on $\beta_1, \ldots, \beta_q$) is $1$-summable for $p = q+2$, $q > 0$.

# Chapter 4

# Cauchy-Heine Transform

If $\psi(w)$ is a continuous function on the straight line segment from $0$ to a point $a$ $(\neq 0)$, then

$$f(z) = \frac{1}{2\pi i} \int_0^a \psi(w)(w-z)^{-1}dw$$

will be analytic for $z$ in the complex plane with a "cut" from $0$ to $a$, but generally, $f(z)$ will be singular at the origin (and at $z = a$).

We will see that $f(z)$ will have an asymptotic power series expansion (of Gevrey order $k > 0$) at the origin, provided $\psi(w)$ is "asymptotically zero" (of order $k$) as $w \to 0$. Hence integrals of the above type provide an excellent tool for making up examples of functions with asymptotic expansions or of series which are $k$-summable in certain directions; this will become clearer very soon.

## 4.1 Definition and Basic Properties

If $k > 0$ and a sector $S$ are given, we write

$$A_k^{(0)}(S)$$

for the set of $\psi \in A_k(S)$ with $J(\psi) = \hat{0}$, i.e. the set of analytic functions $\psi$ (in $S$), such that to every closed subsector $\overline{S_1}$ of $S$ there exist $c_1, c_2 > 0$ for which

$$|\psi(z)| \le c_1 \exp\{-c_2|z|^{-k}\}, \quad z \in \overline{S_1}$$

(compare Ex. 3, Section 2.2). Let $\psi \in A_k^{(0)}(S)$, $S = S(d, \alpha, \rho)$, and fix $a \in S$. Then the function

$$f(z) = CH_a(\psi)(z) = \frac{1}{2\pi i} \int_0^a \psi(w)(w-z)^{-1}dw$$

will be called *Cauchy-Heine Transform* of $\psi(w)$ (with integration along the straight line segment). Clearly, $f(z)$ is analytic for $z$ (on the Riemann surface of the Logarithm) with

$$\arg a < \arg z < 2\pi + \arg a \,,$$

and vanishes as $z \to \infty$ ($f(z)$ even is analytic at $\infty$, if we consider $z$ in the complex plane instead of the Riemann surface, but that is of no importance right now). By deforming the path of integration, we can analytically continue $f(z)$ into the sector $\tilde{S} = S(\tilde{d}, \tilde{\alpha}, \tilde{\rho})$, with $\tilde{d} = d + \pi$, $\tilde{\alpha} = \alpha + 2\pi$, $\tilde{\rho} = |a|$.

**Proposition 1.** *Let* $S, \psi, \tilde{S}, f$ *be as above. Then*

$$f(z) \cong_k \hat{f}(z) \quad in \quad \tilde{S},$$

*with* $\hat{f}(z) = \sum_0^\infty f_n z^n$ *so that*

$$f_n = \frac{1}{2\pi i} \int_0^a \psi(w) w^{-n-1} dw, \quad n \geq 0.$$

*Moreover, if both* $z$ *and* $ze^{2\pi i}$ *are in* $\tilde{S}$ *(i.e.* $z \in S$ *and* $|z| < |a|$*), then*

$$f(z) - f(ze^{2\pi i}) = \psi(z).$$

**Proof.** For $N \geq 0$,

$$(w - z)^{-1} = \sum_{n=0}^{N-1} z^n w^{-n-1} + z^N w^{-N} (w - z)^{-1},$$

hence (with $f_n$ as above)

$$r_f(z, N) = \frac{1}{2\pi i} \int_0^a \psi(w) w^{-N} (w - z)^{-1} dw, \quad z \in \tilde{S},$$

(if we integrate according to $z$). For each closed subsector $\overline{S_1}$ of $\tilde{S}$ of opening less than $2\pi$ (larger subsectors can be split into finitely many peaces), one can choose a path of integration from $0$ to $a$, so that $c = c(\overline{S_1}) > 0$ exists for which

$$|w - z| \geq c|w|,$$

for every $w$ on the path and every $z \in \overline{S_1}$. Since $\psi \in A_k^{(0)}(S)$, we have for sufficiently large $C, K > 0$ (independent of $w$)

$$|w^{-N} \psi(w)| \leq C K^N \Gamma(1 + N/k),$$

for every $N \geq 0$ and every $w$ on the path of integration. This implies (with $L$ being the length of the path of integration)

$$|r_f(z, N)| \leq c^{-1} CL(2\pi)^{-1} K^{N+1} \Gamma(1 + (N+1)/k),$$

for every $N \geq 0$ and $z \in \overline{S_1}$, from which follows $f(z) \cong_k \hat{f}(z)$ in $\tilde{S}$. To prove the remaining identity, we observe

$$f(z) - f(ze^{2\pi i}) = \frac{1}{2\pi i} \oint_\gamma \psi(w)(w - z)^{-1} dw$$

with a closed path of positive orientation around $z$, hence Cauchy's Formula completes the proof. $\square$

**Remarks.**

1. Under the assumptions of Proposition 1, we see immediately that $\hat{f}(z)$ is $k$-summable in all directions $\hat{d}$ with

$$|\hat{d} - d - \pi| < \alpha/2 + \pi(1 - 1/(2k))$$

(provided that the right hand side is positive). Since $\alpha$ may be arbitrarily small, it may happen for $0 < k < 1/2$ that there is no such $\hat{d}$. We will, later on, have more evidence for the fact that values of $k$ below (or even equal to) $1/2$ are special.

2. It is worthwhile to observe for later applications that for $a, b \in S$ (and $\psi \in A_k^{(0)}(S)$) the function $CH_a(\psi) - CH_b(\psi) = g$ is *analytic at the origin*.

3. The formal series $\hat{f}$ in Proposition 1 will sometimes be denoted as $\widehat{CH}_a(\psi)$ (*formal Cauchy-Heine Transform*).

## 4.2 Normal Coverings

Although we sometimes consider sectors of opening larger than $2\pi$ (and therefore have to work on the Riemann surface of the Logarithm) we have seen in Chapter 3 that for $k$-summability we should identify directions $d$ which differ by integer multiples of $2\pi$ (hence should think of rays $\arg z = d$ in the complex plane). Quite similarly, the notion of a *normal covering* (to be defined) is best understood if pictured in the complex plane:

For a natural number $m$, let

$$d_0 < d_1 < \ldots < d_{m-1} < d_0 + 2\pi =: d_m$$

be given directions. Moreover, let $\alpha_j > 0$ $(j = 0, \ldots, m-1)$ be also given, so that with $\alpha_m := \alpha_0$

$$d_j - \alpha_j/2 < d_{j-1} + \alpha_{j-1}/2, \quad 1 \leq j \leq m .$$

If this is so, and $\rho > 0$ is arbitrarily given, then we say that the sectors

$$S_j = S(d_j, \alpha_j, \rho), \quad 0 \leq j \leq m ,$$

form *a normal covering*. One should observe that (on the Riemann surface) $S_m$ is directly above $S_0$. One can also define $d_j$ and $\alpha_j$, for arbitrary integers $j$, so that

$$d_{j+m} = d_j + 2\pi , \quad \alpha_{j+m} = \alpha_j \quad \text{for every} \quad j ;$$

then the sectors $S_j = S(d_j, \alpha_j, \rho)$ cover the whole Riemann surface, and $m + 1$ consecutive ones always are a normal covering.

**Theorem 1.** *Let* $S_j = S(d_j, \alpha_j, \rho)$, $0 \le j \le m$, *be a normal covering. For* $k > 0$ *and* $1 \le j \le m$, *let*

$$\psi_j \in A_k^{(0)}(S_{j-1} \cap S_j),$$

*and choose* $a_j \in S_{j-1} \cap S_j$. *For* $0 < \tilde{\rho} \le |a_j|$ $(1 \le j \le m)$, *let* $\tilde{S}_j = S(d_j, \alpha_j, \tilde{\rho})$, $0 \le j \le m$, *and define*

$$f_j(z) = \sum_{\mu=1}^{j} CH_{a_\mu}(\psi_\mu)(z) + \sum_{\mu=j+1}^{m} CH_{a_\mu}(\psi_\mu)(ze^{2\pi i})$$

*for* $z \in \tilde{S}_j$, $0 \le j \le m$ *(interpreting empty sums as zero). Then for* $j = 0, \dots, m$

$$f_j \in A_k(\tilde{S}_j),$$

*with*

$$J(f_j) = \sum_{\mu=1}^{m} \widehat{CH}_{a_\mu}(\psi_\mu),$$

*(hence* $J(f_j)$ *is independent of* $j$). *Moreover,*

$$
\begin{aligned}
f_j(z) - f_{j-1}(z) &= \psi_j(z), & 1 \le j \le m, \\
f_m(ze^{2\pi i}) &= f_0(z), & z \in S_0.
\end{aligned}
$$

**Proof.** For each $\mu$, $CH_{a_\mu}(\psi_\mu)$ is defined and analytic in $S(d_\mu + \pi, \alpha + 2\pi, \tilde{\rho})$; so one easily checks for $1 \le \mu \le j$ (resp. $j + 1 \le \mu \le n$) that $CH_{a_\mu}(\psi_\mu)(z)$ (resp. $CH_{a_\mu}(\psi_\mu)(ze^{2\pi i})$) are analytic in $\tilde{S}_j$ $(0 \le j \le m)$. The proof then is easily completed, using Proposition 1.      □

## 4.3   Decomposition Theorems

For $\psi \in A_k^{(0)}(S)$, the coefficients of $\hat{f} = \widehat{CH}_a(\psi)$ (for arbitrary $a \in S$) have the form of a *moment sequence*, so that $\hat{f}$ appears, at first glance, to be rather special. The following Theorem shows, however, that arbitrary $\hat{f} \in \mathbb{C}[[z]]_{1/k}$ admit a decomposition into a convergent series plus finitely many formal Cauchy-Heine Transforms:

**Theorem 2.** *For* $k > 0$ *and* $\hat{f} \in \mathbb{C}[[z]]$, *the following statements are equivalent:*

a)            $\hat{f} \in \mathbb{C}[[z]]_{1/k}$ .

b) *For every normal covering* $S_j = S(d_j, \alpha_j, \rho)$ *with* $0 < \alpha_j \le \pi/k$, $0 \le j \le m$, *there exist* $\psi_j \in A_k^{(0)}(S_{j-1} \cap S_j)$, $1 \le j \le m$, *so that (for arbitrarily chosen* $a_j \in S_{j-1} \cap S_j$)

$$\hat{f} = \hat{f}_0 + \sum_{j=1}^{m} \widehat{CH}_{a_j}(\psi_j), \quad \text{with} \quad \hat{f}_0 \in \mathbb{C}\{z\}.$$

**Proof.** Since b) trivially implies a), let now $\hat{f} \in \mathbb{C}[[z]]_{1/k}$ and let $S_j$ be as stated, $0 \le j \le m$. Then Proposition 1, Section 2.2, implies existence of $\tilde{f}_j \in A_k(S_j)$ with $J(\tilde{f}_j) = \hat{f}$, for $j = 0, \ldots, m-1$, and defining $\tilde{f}_m(z) = \tilde{f}_0(ze^{-2\pi i})$, $z \in S_m$, the same holds for $j = m$. Let

$$\psi_j(z) = \tilde{f}_j(z) - \tilde{f}_{j-1}(z), \quad z \in S_{j-1} \cap S_j ,$$

then $\psi_j \in A_k^{(0)}(S_{j-1} \cap S_j)$, $1 \le j \le m$. With $f_j$ as in Theorem 1, we then see that

$$f(z) = \tilde{f}_j(z) - f_j(z)$$

is independent of $j$, and analytic and single-valued in a punctured neighborhood of the origin. Moreover, $f$ has an asymptotic expansion there, hence Proposition 1 b), Section 1.4, implies analyticity of $f$ at the origin. From

$$\tilde{f}_j(z) = f(z) + f_j(z)$$

we conclude

$$\hat{f} = J(\tilde{f}_j) = \hat{f}_0 + \sum_{\mu=1}^{m} \widehat{CH}_{a_\mu}(\psi_\mu) ,$$

with convergent $\hat{f}_0$. $\qquad\qquad\qquad\qquad\qquad\qquad\qquad\qquad\qquad\qquad$ $\square$

**Remark.** Theorem 2 (and its proof) remain correct if in b) we allow normal coverings (of possibly larger openings) to which $f_j \in A_k(S_j)$ with $J(f_j) = \hat{f}$, $0 \le j \le m-1$, still exist. This will be used in the proof of the following

**Corollary to Theorem 2.** *Let* $k > 0$, $d \in \mathbb{R}$ *and* $\hat{f} \in \mathbb{C}[[z]]$ *be given.*

a) *For* $k \ge 1/2$, $k$*-summability of* $\hat{f}$ *in direction* $d$ *is equivalent to the existence of a decomposition of* $\hat{f}$ *as in Theorem 2 b, with*

$$|d + \pi - \arg a_j| \le \pi - \pi/(2k), \quad 1 \le j \le m .$$

b) *For* $k \le 1/2$, $k$*-summability of* $\hat{f}$ *in direction* $d$ *is equivalent to the existence of a decomposition of* $\hat{f}$ *as in Theorem 2 b, with* $m = 1$ *and*

$$S_0 \cap S_1 = S(d + \pi, \alpha, \rho), \quad \alpha > \pi/k - 2\pi .$$

**Proof.** Let $\hat{f} \in \mathbb{C}\{z\}_{k,d}$ and $k \ge 1/2$, then we can, according to the above remark, take a normal covering $S_j$, $j = 0, \ldots, m$, where $S_0$ has bisecting direction $d$ and opening larger than $\pi/k$. Since we are free to choose any $a_j \in S_{j-1} \cap S_j$, it is easy to see that we can arrange $|d + \pi - \arg a_j| \le \pi - \pi/(2k)$. In case $0 < k \le 1/2$, a single sector of opening larger than $\pi/k$ already forms a normal covering, and $\hat{f} \in \mathbb{C}\{z\}_{k,d}$ implies $f = S_{k,d}(\hat{f}) \in A_k(S_0)$, $S_0 = S(d, \tilde{\alpha}, \rho)$, $\tilde{\alpha} > \pi/k$. With $S_1 = S(d + 2\pi, \tilde{\alpha}, \rho)$ we have that $S_0 \cap S_1$ is a sector of opening larger than $\pi/k - 2\pi$ and bisecting direction $d + \pi$. So one direction of the proof (in both cases) is completed. The opposite direction, however, follows from Remark 1, Section 4.1. $\qquad$ $\square$

**Theorem 3.**  *Let*  $k > 0$  *and*  $\hat{f} \in \mathbb{C}\{z\}_k$  *be given, and let*  $\hat{f}$  *have*  $m \geq 2$  *singular directions (modulo*  $2\pi$ *). Then*

$$\hat{f} = \hat{f}_1 + \ldots + \hat{f}_m \, ,$$

*where each*  $\hat{f}_j \in \mathbb{C}\{z\}_k$  *has exactly one singular direction.*

**Proof.**  Let the singular directions of  $\hat{f}$  be

$$0 < d_1 < \ldots < d_m \leq 2\pi \, ,$$

and define  $d_0 = d_m - 2\pi$ . For each  $j$ ,  $1 \leq j \leq m$ , define

$$f_j = S_{k,d}(\hat{f}) \, , \quad d_{j-1} < d < d_j \, ,$$

then  $f_j$  is analytic in some region  $G_j$  such that to every  $\varepsilon > 0$  there exists a sector in  $G_j$  with bisecting direction  $(d_{j-1}+d_j)/2$  and opening  $d_j - d_{j-1}+\pi/k-\varepsilon$  (and sufficiently small radius), and in each such sector,  $f_j(z) \cong_k \hat{f}(z)$ . With  $G_0 = G_m e^{-2\pi i}$ ,  $f_0(z) = f_m(ze^{2\pi i})$ ,  $z \in G_0$ , define

$$\psi_j(z) = f_j(z) - f_{j-1}(z) \, , \quad z \in G_{j-1} \cap G_j \, , \quad 1 \leq j \leq m \, ,$$

and choose  $a_j \in G_{j-1} \cap G_j$  arbitrarily. It follows from above that to every  $\varepsilon > 0$  there exists a sector in  $G_{j-1} \cap G_j$  with bisecting direction  $d_{j-1}$  and opening  $\pi/k-\varepsilon$ . Using this, one can see from Remarks 1 and 2, Section 4.1, that  $\widehat{CH}_{a_j}(\psi_j)$  is  $k$ -summable in all directions but  $d_{j-1}$  (modulo  $2\pi$ ). As in the proof of Theorem 2 one can show that  $\hat{f} - \sum_{j=1}^{m} \widehat{CH}_{a_j}(\psi_j) = \hat{f}_0$  converges (hence is also  $k$ -summable). Defining  $\hat{f}_j = \widehat{CH}_{a_j}(\psi_j)$ ,  $1 \leq j \leq m-1$ , and  $\hat{f}_m = \hat{f}_0 + \widehat{CH}_{a_m}(\psi_m)$ , the proof is completed.                                                                                    □

## 4.4    A Characterization of Functions with Gevrey Asymptotic

The following result can be thought of as a generalization of the well-known characterization of removable singularities. As we shall see in the exercises below, it is very useful in showing that some functions have a Gevrey asymptotic.

**Proposition 2.**  *Let*  $k > 0$ , *any sector*  $S$  *and any function*  $f$ , *analytic in*  $S$ , *be given. Then*  $f \in A_k(S)$  *is equivalent to the existence of a normal covering*  $S_0, \ldots, S_m$ , *with*  $S_0 = S$ , *and functions*  $f_j$ , *analytic in*  $S_j$   $(0 \leq j \leq m)$ , *with*  $f_0 = f$  *and*  $f_m(z) = f_0(ze^{-2\pi i})$ ,  $z \in S_m$ , *so that all*  $f_j$  *are bounded at the origin, and*

$$f_{k-1}(z) - f_k(z) \in A_k^{(0)}(S_{k-1} \cap S_k) \, , \quad 1 \leq k \leq m \, .$$

**Proof.** If $f \in A_k(S)$, let $\hat{f} = J(f)$, and choose any normal covering $S_0, \ldots, S_m$ so that $S_0 = S$ and the opening of $S_1, \ldots, S_{m-1}$ is less than or equal to $\pi/k$. Then Proposition 1, Section 2.2, shows existence of $f_j \in A_k(S_j)$ with $J(f_j) = \hat{f}$, $1 \le j \le m-1$, and with $f_m(z) = f_0(ze^{-2\pi i})$ we obtain $f_{k-1}(z) - f_k(z) \in A_k^{(0)}(S_{k-1} \cap S_k)$, $1 \le k \le m$. Conversely, if $S_j$ and $f_j$ are as stated, define

$$\psi_j = f_j - f_{j-1}, \quad 1 \le j \le m,$$

and (with $a_\mu \in S_{\mu-1} \cap S_\mu$ and $\tilde{S}_j$ as in Theorem 1)

$$g_j(z) = \sum_{\mu=1}^{j} C H_{a_\mu}(\psi_\mu)(z) + \sum_{\mu=j+1}^{m} C H_{a_\mu}(\psi_\mu)(ze^{2\pi i})$$

for $z \in \tilde{S}_j$ and $0 \le j \le m$. Then Theorem 1 shows $g_j \in A_k(\tilde{S}_j)$ and $J(g_j) = \hat{g}$, independent of $j$. Moreover, $h(z) = f_j(z) - g_j(z)$ is also independent of $j$, and single-valued and bounded at the origin. Therefore, the origin is a removable singularity of $h$, and consequently $f_j = h + g_j \in A_k(\tilde{S}_j)$, and even is in $A_k(S_j)$, because $\tilde{S}_j$ and $S_j$ only differ in the radius (compare Ex. 1, Section 1.5).

**Exercises.** Throughout the following, let $\lambda$ be any complex number, let $p(z)$ be a polynomial of degree $r \ge 1$ and highest coefficient one, and let $g(u)$ be analytic (and single-valued) for $0 \le |u| < \rho$.

1. For $j = 0, \ldots, r$, define

$$f_j(1/z) = z^\lambda e^{p(z)} \int_{\infty(2j\pi/r)}^{z} u^{-\lambda} e^{-p(u)} g(1/u) du, \quad 0 < |z| < \rho,$$

integrating from $\infty$ along the line $\arg u = 2j\pi/r$ to some point $z_{0,j}, |z_{0,j}|^{-1} < \rho$, and then to $z$. Show that each $f_j$ is analytic (on the Riemann surface of the Logarithm) for $0 < |z| < \rho$, and

$$f_j(1/z) - f_{j-1}(1/z) = c_j z^\lambda e^{p(z)}, \quad j = 1, \ldots, r,$$

with

$$c_j = \int_{\gamma_j} u^{-\lambda} e^{-p(u)} g(1/u) du,$$

where $\gamma_j$ is a path from $\infty$ along $\arg u = 2j\pi/r$ to $z_{0,j}$, then to $z_{0,j-1}$, and back to $\infty$ along $\arg u = 2(j-1)\pi/r$.

2. Show that $f_j(z)$ is bounded at the origin, for $z \in S_j = S(2j\pi/r, 3\pi/r, \rho)$, and $j = 0, \ldots, r$.

*Hint.* Use a change of variable in the integral to reduce the cases $j = 1, \ldots, r$ to a case $j = 0$, but with $p(z)$ and $g(u)$ depending upon $j$.

3. Show
$$f_{j-1}(z) - f_j(z) \in A_r^{(0)}(S_{j-1} \cap S_j), \quad 1 \leq j \leq r,$$
$$f_r(z) = f_0(ze^{-2\pi i}), \quad z \in S_r.$$

4. Show  $f_j \in A_r(S_j), \ 0 \leq j \leq r.$

# Chapter 5

# Acceleration Operators

Let $f$ be a function, analytic in a sector $S$ and bounded at the origin. Let $\tilde{k} > k > 0$ be such that $g_k = \mathcal{B}_k f$ and $g_{\tilde{k}} = \mathcal{B}_{\tilde{k}} f$ both exist. Then we may wish to know whether or not $g_{\tilde{k}}$ can be related to $g_k$ by means of an integral transformation, and if so, what the kernel in this transformation looks like. This question leads to the definition of so-called *acceleration operators*, which were first introduced by J. Ecalle. Like the Laplace operator, they will turn out to be bijective operators (between certain differential algebras), and the inverse operators will briefly be discussed in a number of exercises in Section 5.2.

The acceleration operators will be the essential tool in our treatment of multisummability in the remaining chapters.

## 5.1    Definition and Properties of the Kernel

For real $\alpha > 1$ and complex $t$, let

$$C_\alpha(t) = \frac{1}{2\pi i} \int_\gamma u^{1/\alpha - 1} \exp\{u - tu^{1/\alpha}\} du \,,$$

with a path of integration $\gamma$ as in Hankel's integral for the inverse Gamma function: from $\infty$ along $\arg u = -\pi$ to some $u_0 < 0$, then on the circle $|u| = |u_0|$ to $\arg u = \pi$, and back to $\infty$ along this ray. Because of $\alpha > 1$, this integral represents an entire function of $z$.

**Lemma 1.** *Let $\alpha > 1$, and $\beta$ be the conjugate index, i.e.*

$$1/\beta + 1/\alpha = 1 \,.$$

*Then to every $\varepsilon > 0$, there exist $c_1, c_2 > 0$ so that*

$$|C_\alpha(t)| \le c_1 \exp\{-c_2 |t|^\beta\}$$

*for every $t$ with $|\arg t| \le \pi/(2\beta) - \varepsilon$.*

**Proof.**  Make a change of variable $tu^{1/\alpha} = z^{-1}$ in the integral representation, and then replace $t$ by $t^{-1}$ to see that $t^{-1}C_\alpha(t^{-1})$ is the Borel Transform of index $\alpha$ of $z^{-1}e^{-1/z}$. From Theorem 2, Section 2.3, we then obtain $C_\alpha(t^{-1}) \in A_\beta^{(0)}(S)$, for $S = S(0, \pi/\beta, +\infty)$, and Ex. 3, Section 2.2 can be used to complete the proof.  $\square$

**Remark.**  One can considerably strengthen Lemma 1, e.g. by explicitly giving the values of $c_1, c_2$ (depending upon $\varepsilon$), but we need not do this here.

For real numbers $d$ and $\tilde{k} > k > 0$, we now define

$$(A_{\tilde{k},k}f)(z) = z^{-k} \int\limits_0^{\infty(d)} f(t)C_\alpha\left((t/z)^k\right) d(t^k)$$

(with $\alpha = \tilde{k}/k > 1$), whenever $f$ is so that the integral gives sense, at least for some $z$ . We shall call $A_{\tilde{k},k}$ the *acceleration operator* (with indices $\tilde{k}$ and $k$, bearing in mind that $\tilde{k} > k$ always holds). Our definition of acceleration operators differs slightly from the one in the original papers of Ecalle, Martinet-Ramis and others; this is so to make them match with our definition of Borel and Laplace Transform.

**Exercises.**

1. Compute the power series expansion of $C_\alpha(t)$, $\alpha > 1$, and show for every fixed $t$

$$\lim_{\alpha \to \infty} C_\alpha(t) = e^{-t} .$$

2. For $\tilde{k} > k > 0$ and $\alpha = \tilde{k}/k$, show

$$\int\limits_0^{\infty(d)} t^{-k}C_\alpha\left((u/t)^k\right) e^{-(t/z)^{\tilde{k}}} d(t^{\tilde{k}}) = z^{\tilde{k}-k} \exp\{-(u/z)^k\} ,$$

for every real number $d$ and every $u$ and $z$ with

$$k|\arg u - d| < \pi(1 - 1/\alpha)/2, \quad |\arg z - d| < \pi/(2\tilde{k}) .$$

3. For $\tilde{k} > k > 0$ and $\alpha = \tilde{k}/k$, let $f$ be analytic in a sector $S = S(d, \alpha)$ of infinite radius, and assume that $f$ is bounded at the origin and of exponential size not more than $k$. Show that

$$z^{-k} \int\limits_0^{\infty(d)} f(u) \exp\{-(u/z)^k)\} d(u^k) =$$

$$z^{-\tilde{k}} \int\limits_0^{\infty(d)} t^{-k} \left[ \int\limits_0^{\infty(d)} f(u)C_\alpha\left((u/t)^k\right) d(u^k) \right] \exp\{-(t/z)^{\tilde{k}}\} d(t^{\tilde{k}})$$

for every $z = re^{i\varphi}$ with

$$-\pi/2 < \tilde{k}(\varphi - d) < \pi/2, \quad \cos\tilde{k}(\varphi - d) > cr^{\tilde{k}},$$

where $c > 0$ is sufficiently large.

*Hint.* Use Lemma 1 to show absolute convergence of the double integral on the right, then interchange the order of integration, and use Ex. 2.

4. Use Ex. 3 to show for $f(u) = u^{\lambda}$ $(\lambda \in \mathbb{C}, \text{ Re }\lambda \geq 0)$ that

$$(\mathcal{A}_{\tilde{k},k}f)(z) = z^{\lambda}\frac{\Gamma(1 + \lambda/k)}{\Gamma(1 + \lambda/\tilde{k})},$$

provided $\text{Re }\lambda \geq 0$.

In view of Ex. 4, it is natural to define for $\hat{f}(z) = \sum_{0}^{\infty} f_n z^n$

$$(\hat{\mathcal{A}}_{\tilde{k},k}\hat{f})(z) = \sum_{0}^{\infty} z^n f_n \frac{\Gamma(1 + n/k)}{\Gamma(1 + n/\tilde{k})}$$

(the *formal acceleration operator*).

5. Let

$$f(z) = \sum_{n=0}^{\infty} f_n z^n / \Gamma(1 + n/\kappa)$$

with $\kappa > 0$ and $|f_n| \leq c^n$, $n \geq 0$ (for some fixed $c > 0$). Recall from Section 1.2 that then $f(z)$ is an entire function of exponential size at most $\kappa$ (in every sector of infinite radius). For $\tilde{k} > k > 0$ so that

$$1/\kappa = 1/k - 1/\tilde{k}$$

(and arbitrary $d$), show that $\mathcal{A}_{\tilde{k},k}f$ is an analytic function near the origin, and compute its power series expansion.

*Hint.* Show that termwise integration of the power series expansion of $f$ is justified.

## 5.2 Basic Properties of the Acceleration Operators

The proof of the following theorem proceeds along the same lines as that of Theorem 1 in Section 2.1 and will be left to the reader.

**Theorem 1.**   *Let  $S = S(d, \alpha)$  be a sector of infinite radius, and let  $\tilde{k} > k > 0$  and  $k_1 > 0$  be given. Let  $f$  be analytic and of exponential size not more than  $\kappa$  in  $S$,  with*

$$1/\kappa = 1/k - 1/\tilde{k} ,$$

*and let*

$$f(z) \cong_{k_1} \hat{f}(z) \quad in \quad S .$$

*Then to every  $\varepsilon > 0$  there exists  $\rho = \rho(\varepsilon) > 0$  so that  $g = A_{\tilde{k}, k} f$  is analytic in the sector*

$$\tilde{S} = S(d, \alpha + \pi/\kappa - \varepsilon, \rho) ,$$

*and*

$$g(z) \cong_{k_2} \hat{g}(z) \quad in \quad \tilde{S} ,$$

*with*

$$\hat{g} = \hat{A}_{\tilde{k}, k} \hat{f} \quad and \quad 1/k_2 = 1/k_1 + 1/\kappa .$$

The proof of the next theorem can be seen to follow from the previous one and Ex. 3 in Section 5.1:

**Theorem 2.**   *In addition to the assumptions of Theorem 1, let  $f$  be of exponential size not more than  $k$.  Then  $g = A_{\tilde{k}, k} f$  is analytic and of exponential size not more than  $\tilde{k}$  in  $\tilde{S} = S(d, \alpha + \pi/\kappa)$,  and*

$$\mathcal{L}_{\tilde{k}} g = \mathcal{L}_k f .$$

**Exercises.**

1. For  $\tilde{k} > \hat{k} > k > 0$,  show

$$z^{-k} C_{\hat{k}/k} \left( (u/z)^k \right) = z^{-k} \int_0^{\infty(d)} t^{-k} C_{\tilde{k}/k} \left( (t/z)^{\tilde{k}} \right) C_{\hat{k}/k} \left( (u/t)^k \right) d(t^{\tilde{k}}) ,$$

   for every real number  $d$,  and every  $u$  and  $z$  with

$$|\arg u - d| < \pi(1/k - 1/\hat{k})/2, \quad |\arg z - d| < \pi(1/\hat{k} - 1/\tilde{k})/2 .$$

   **Hint.** For fixed  $u$,  let  $f(t) = t^{-k} C_{\hat{k}/k}((u/t)^k)$,   $g(z) = z^{-k} C_{\tilde{k}/k}((u/z)^k)$,  and use Ex. 2 from the previous section to see  $\mathcal{L}_{\tilde{k}} g = \mathcal{L}_{\hat{k}} f$;  then use Theorem 2 to conclude  $g = A_{\tilde{k}, \hat{k}} f$.

2. Under the assumptions of Theorem 1, let  $\hat{k}$  be in the interval  $(k, \tilde{k})$.

   a) Show that  $\tilde{g} = A_{\hat{k}, k} f$  is analytic and of exponential size not more than  $\hat{\kappa}$,   $1/\hat{\kappa} = 1/\hat{k} - 1/\tilde{k}$,  in  $\tilde{S} = S(d, \alpha + \pi/\tilde{\kappa})$,   $1/\tilde{\kappa} = 1/k - 1/\hat{k}$.

b) Show $A_{\tilde{k},k}f = A_{\tilde{k},\tilde{k}}\tilde{g}$, where integration in $A_{\tilde{k},\tilde{k}}$ can be along any ray in $\tilde{S}$.

*Hint.* For b), use Ex. 1.

In the following exercises, let

$$D_\alpha(z) = \sum_{n=0}^\infty z^n \frac{\Gamma(1+n/\alpha)}{n!}, \quad z \in \mathbb{C} \quad \alpha > 1.$$

3.  a) Show that $D_\alpha$ is an entire function of exponential size $\beta = \alpha/(\alpha-1)$, and

$$D_\alpha(z) = \int_0^{\infty(d)} \exp\left[zx^{1/\alpha} - x\right] dx ,$$

for every $z$, and $d \in (-\pi/2, \pi/2)$.

b) Make a change of variable $x^{1/\alpha} = t$ to show

$$D_\alpha(z) = \alpha \int_0^{\infty(d)} t^{\alpha-1} e^{-t^\alpha} e^{zt} dt ,$$

for every $z$, and $d \in (-\pi/(2\alpha), \pi/(2\alpha))$. Use this to show

$$D_\alpha(-1/z) \cong_1 \alpha \sum_{n=0}^\infty (-1)^n z^{\alpha(n+1)} \frac{\Gamma(\alpha(n+1))}{n!}$$

in $S(0, \pi(1+1/\alpha))$. (Here we use a more general definition of Gevrey asymptotics: For $\alpha, k > 0$, we say that

$$f(z) \cong_k \sum_{n=0}^\infty f_n z^{\alpha n} \quad \text{in} \quad S(d, \beta, r)$$

iff

$$f(z^{1/\alpha}) \cong_{k/\alpha} \sum_{n=0}^\infty f_n z^n \quad \text{in} \quad S(\alpha d, \alpha\beta, r^\alpha) .$$

Compare with Ex. 2 in Section 1.5 to see that this is correct in case of $\alpha$ being a natural number.)

4. For $\tilde{k} > k > 0$, $\alpha = \tilde{k}/k$, show

$$z^{-k} \int_0^{\infty(d)} C_\alpha((t/z)^k) D_\alpha((t/u)^k) d(t^k) = \frac{u^k}{u^k - z^k} ,$$

for arbitrary real $d$, and $z, u$ with

$$|d - \arg z| < \pi(1/k - 1/\tilde{k})/2, \quad |z|, |u| \text{ sufficiently small}.$$

*Hint.* Check that termwise integration of the expansion for $D_\alpha((t/u)^k)$ is justified for sufficiently small $|z/u|$.

In the remaining exercises, fix $\tilde{k} > k > 0$, $k_1 > 0$ and a sector $S = S(d, \alpha, \rho)$ of opening $\alpha > \pi/\kappa$, where $1/\kappa = 1/k - 1/\tilde{k}$. For $f \in A_{k_1}(S)$, define (with $\alpha = \tilde{k}/k > 1$)

$$\mathcal{D}_{\tilde{k}, k}(f)(u) = \frac{1}{2\pi i} \int_{\gamma_\kappa(\tau)} z^k f(z) D_\alpha((u/z)^k) d(z^{-k}),$$

with $\gamma_\kappa(\tau)$ as in Section 2.3. The operator $\mathcal{D}_{\tilde{k}, k}$ is called *deceleration operator* (with indices $\tilde{k}$ and $k$).

5. Show that for $f \in A_{k_1}(S)$, the above integral converges absolutely and compactly for $u \in S(d, \alpha - \pi/\kappa, +\infty)$, thus $\mathcal{D}_{\tilde{k}, k}(f) = g$ is analytic in $\tilde{S} = S(d, \alpha - \pi/\kappa, +\infty)$. Moreover, show that $g$ is of exponential size at most $\kappa$ in $\tilde{S}$.

6. For $f(z) = z^\lambda$ ($\lambda \in \mathbb{C}$, $\text{Re}\,\lambda \geq 0$), show that

$$\mathcal{D}_{\tilde{k}, k}(f)(u) = u^\lambda \frac{\Gamma(1 + \lambda/\tilde{k})}{\Gamma(1 + \lambda/k)}.$$

7. Show that (for $f \in A_{k_1}(S)$)

$$g(z) = \mathcal{D}_{\tilde{k}, k}(f)(z) \cong_{k_2} \hat{g}(z) \quad \text{in} \quad \tilde{S},$$

with

$$\hat{g}(z) = \sum_{n=0}^{\infty} f_n z^n \frac{\Gamma(1 + n/\tilde{k})}{\Gamma(1 + n/k)}, \quad \hat{f}(z) = \sum_{n=0}^{\infty} f_n z^n = J(f),$$

and

$$1/k_2 = 1/k_1 - 1/\kappa \quad \text{if} \quad \kappa > k_1, \quad \text{resp.} \quad k_2 = \infty \quad \text{if} \quad \kappa \leq k_1.$$

8. Show for $f \in A_{k_1}(S)$

$$A_{\tilde{k}, k} \circ \mathcal{D}_{\tilde{k}, k} f = f,$$

and for $g \in A_{k_2}(\tilde{S})$

$$\mathcal{D}_{\tilde{k}, k} \circ A_{\tilde{k}, k} g = g.$$

## 5.3 Convolutions

Let $f$ and $g$ be analytic in a sector $S$, and so that

$$f(z) \quad \cong_{k_1} \quad \hat{f}(z) \quad \text{in} \quad S,$$

$$g(z) \quad \cong_{k_1} \quad \hat{g}(z) \quad \text{in} \quad S,$$

for some $k_1 > 0$. For arbitrary $k > 0$, we then define an analytic function $h(z)$ (for $z \in S$) by

$$h(z^{1/k}) = \frac{d}{dz} \left[ \int_0^z f\left((z-t)^{1/k}\right) g(t^{1/k}) dt \right].$$

We call $h$ the convolution of $f$ and $g$ (with index $k$), and write in short-hand

$$h = f *_k g.$$

**Lemma 2.** *For $f, g, h, S, k$ and $k_1$ as above, we have*

$$h(z) \cong_{k_1} \hat{h}(z) \quad (=: \hat{f} *_k \hat{g}),$$

*where $\hat{h}(z) = \sum_0^\infty h_n z^n$, with*

$$h_n \Gamma(1 + n/k) = \sum_{m=0}^n f_{n-m} \Gamma(1 + (n-m)/k) g_m \Gamma(1 + m/k), \quad n \geq 0.$$

*In case $S$ is of infinite radius and $f, g$ are of exponential size not more than $\kappa$ (with some $\kappa > 0$), then so is $h(z)$.*

**Proof.** By assumption, we have for every closed subsector $\hat{S}$ of $S$ and sufficiently large $C, K > 0$

$$|r_f(z, N)| \quad \leq \quad CK^N \Gamma(1 + N/k_1),$$

$$|r_g(z, N)| \quad \leq \quad CK^N \Gamma(1 + N/k_1),$$

for every $N \geq 0$ and $z \in \hat{S}$. Using the Beta-Integral, one can show for $z^{1/k} \in \hat{S}$

$$z^{N/k} r_h(z^{1/k}, N) = \frac{d}{dz} \int_0^z \left[ f\left((z-t)^{1/k}\right) t^{N/k} r_g(t^{1/k}, N) \right.$$

$$\left. + \sum_{m=0}^{N-1} g_m t^{m/k} (z-t)^{(N-m)/k} r_f\left((z-t)^{1/k}, N-m\right) \right] dt.$$

Using the above estimates, we find

$$\left| \int_0^z f\left((z-t)^{1/k}\right) t^{N/k} r_g(t^{1/k}, N) dt \right| \leq C^2 K^N \Gamma(2 + N/k_1) |z|^{1+N/k},$$

$$\left| \int_0^z \sum_{m=0}^{N-1} g_m t^{m/k}(z-t)^{(N-m)/k} \, r_f\left((z-t)^{1/k}, N-m\right) dt \right|$$

$$\leq \frac{C K^N |z|^{1+N/k}}{\Gamma(2+N/k)} \sum_{m=0}^{N-1} \Gamma(1+m/k)\Gamma(1+m/k_1)\Gamma(1+\tfrac{N-m}{k})\Gamma(1+\tfrac{N-m}{k_1})$$

(for $z^{1/k} \in \hat{S}$ and $N \geq 0$). Stirling's Formula implies

$$\sum_{m=0}^{N-1} \Gamma(1+\tfrac{m}{k})\Gamma(1+\tfrac{m}{k_1})\Gamma(1+\tfrac{N-m}{k})\Gamma(1+\tfrac{N-m}{k_1}) \leq \tilde{C}\tilde{K}^N \Gamma(2+\tfrac{N}{k})\Gamma(2+\tfrac{N}{k_1}) \, ,$$

for sufficiently large $\tilde{C}, \tilde{K}$ (independent of $N$), and this shows altogether (for sufficiently large $\hat{C}, \hat{K} > 0$, $z$ as above, and $N \geq 0$)

$$\left| \int_0^z w^{N/k} \, r_h(w^{1/k}, N) dw \right| \leq \hat{C}\hat{K}^N |z|^{1+N/k}\Gamma(2+N/k_1) \, .$$

Using Cauchy's Formula for the first derivative, one can see that this implies $h(z) \overset{\sim}{=}_{k_1}$ $\hat{h}(z)$ in $S$. An elementary estimate can be used to show that $h$ is of the desired exponential size, in case $f$ and $g$ are.  $\qquad\square$

**Theorem 3.** *Let $S$ be a sector of infinite radius, let $\tilde{k} > k > 0$ and $k_1 > 0$ be arbitrarily given, and take*

$$1/\kappa = 1/k - 1/\tilde{k} \, .$$

*Moreover, let $f, g \in A_{k_1}(S)$ be of exponential order not more than $\kappa$. Then*

$$A_{\tilde{k},k}(f *_k g) = (A_{\tilde{k},k}f) *_{\tilde{k}} (A_{\tilde{k},k}g) \, .$$

**Proof.** With $\tilde{f} = A_{\tilde{k},k}f$, $\tilde{g} = A_{\tilde{k},k}g$, $\tilde{h} = \tilde{f} *_{\tilde{k}} \tilde{g}$ we find

$$\tilde{h}(z^{1/\tilde{k}}) = \int_0^{\infty(d)} f(u) \int_0^{\infty(d)} g(w)k(z^{1/\alpha}, w, u)d(w^k)d(u^k) \, ,$$

with

$$k(z^{1/\alpha}, w, u) = \frac{d}{dz} \int_0^z (z-t)^{-1/\alpha} \, C_\alpha\left(\frac{u^k}{(z-t)^{1/\alpha}}\right) t^{-1/\alpha} \, C_\alpha\left(\frac{w^k}{t^{1/\alpha}}\right) dt \, .$$

Fixing $w$ and $u$ and defining

$$k(z) = k(z, w, u), \quad h(z) = z^{-1}C_\alpha\left(\frac{u^k}{z}\right), \quad \ell(z) = z^{-1}C_\alpha\left(\frac{w^k}{z}\right),$$

we see that

$$k = h *_\alpha \ell, \quad h = B_\alpha\left(z^{-1}e^{-u^k/z}\right), \quad \ell = B_\alpha\left(z^{-1}e^{-w^k/z}\right).$$

From the exercises below we find that

$$
\begin{aligned}
k &= B_\alpha\left(z^{-2}e^{-(u^k+w^k)/z}\right) \\
&= B_\alpha\left((u^k+w^k)^{-1}(1+z\frac{d}{dz})z^{-1}e^{-(u^k+w^k)/z}\right) \\
&= (u^k+w^k)^{-1}\left(1+z\frac{d}{dz}\right)z^{-1}C_\alpha\left(\frac{u^k+w^k}{z}\right) \\
&= (u^k+w^k)^{-1}\frac{d}{dz}C_\alpha\left(\frac{u^k+w^k}{z}\right)
\end{aligned}
$$

(for every fixed $u$ and $w$). This implies (with $t = u + w$)

$$
\begin{aligned}
\tilde{h}(z^{1/k}) &= \frac{d}{dz}\int_0^{\infty(kd)} f(u^{1/k})\int_0^{\infty(kd)} g(w^{1/k})C_\alpha\left(\frac{u+w}{z}\right)\frac{dwdu}{u+w} \\
&= \frac{d}{dz}\int_0^{\infty(kd)} t^{-1}C_\alpha(t/z)\int_0^t f\left((t-w)^{1/k}\right)g(w^{1/k})dwdt.
\end{aligned}
$$

Because of

$$t^{-1}\frac{d}{dz}C_\alpha(t/z) = -z^{-1}\frac{d}{dt}C_\alpha(t/z),$$

we may integrate by parts to obtain

$$\tilde{h}(z^{1/k}) = z^{-1}\int_0^{\infty(kd)} C_\alpha(t/z)h(t^{1/k})\,dt,$$

or equivalently

$$\tilde{h}(z) = z^{-k}\int_0^{\infty(d)} h(t)C_\alpha\left((t/z)^k\right)d(t^k). \qquad \square$$

**Exercises.**  Let $S$ be a sector of infinite radius, and let $f, g \in A_{\tilde{k}}(S)$ (for some $\tilde{k} > 0$) be of exponential size at most $k$ (for some $k > 0$).

1. Show that

$$\mathcal{L}_k(f *_k g) = (\mathcal{L}_k f)(\mathcal{L}_k g) .$$

2. Let $\delta$ denote the operator

$$(\delta f)(z) = z \frac{d}{dz} f .$$

Show that

$$\delta(\mathcal{L}_k f) = \mathcal{L}_k(\delta f), \quad \delta(A_{k_1,k_2} f) = A_{k_1,k_2}(\delta f) ,$$

for $1/k_1 + 1/k_2 = 1/k$.

3. For $k, \tilde{k}, \kappa > 0$, let $f, g \in A_{\tilde{k}}(S)$, $S = S(d, \alpha, r)$, and so that

$$(f *_k g)(z) \equiv 1 \quad \text{in} \quad S$$

(implying $f(0) = \lim\limits_{\substack{z \to 0 \\ z \in S}} f(z) \neq 0$). Moreover, assume that $f$ can be analytically continued into $\tilde{S} = S(d, \varepsilon)$ $(\varepsilon > 0)$ and is of exponential size $\kappa$ there. Show the same for $g$.

*Hint.* Show that the series

$$\sum_{\nu=0}^{\infty} g_\nu(z) ,$$

with

$$
\begin{aligned}
g_0(z) &\equiv 1/f(0) , \\
g_{\nu+1}(z^{1/k}) &= \frac{-1}{f(0)} \int_0^z \frac{\partial}{\partial z} f((z-t)^{1/k}) g_\nu(t^{1/k}) dt , \quad \nu \geq 0 ,
\end{aligned}
$$

converges compactly in $\tilde{S}$ to the function $g(z)$; then estimate this series to obtain an exponential estimate for $g(z)$.

## 5.4  Admissible Functions

In the sequel, $q$ will denote a fixed natural number, and $k = (k_1, \ldots, k_q)$, $d = (d_1, \ldots, d_q)$ will be $q$-tupels of real numbers obeying the following restrictions:

a) The numbers $k_j$ will be positive and strictly monotonically decreasing:

$$k_1 > k_2 > \ldots > k_q > 0 \,,$$

and setting $k_0 = +\infty$, we define $\kappa = (\kappa_1, \ldots, \kappa_q)$ by

$$1/\kappa_j = 1/k_j - 1/k_{j-1} \,, \quad 1 \le j \le q \,.$$

b) The numbers $d_j$ are such that

$$|d_j - d_{j-1}| \le \frac{\pi}{2\kappa_j} \,, \quad 2 \le j \le q \,.$$

Given such $k$ and $d$, a function $f$ is said to be *admissible with respect to* $k$ *and* $d$, iff the following holds:

$\alpha$) There exist $\tilde{k} > 0$ and a sector $\tilde{S}$ of infinite radius and bisecting direction $d_q$ so that $f \in A_{\tilde{k}}(\tilde{S})$. Moreover, $f$ is required to be of exponential size not more than $\kappa_q$ in $\tilde{S}$.

$\beta$) In case $q \ge 2$, the function $g = \mathcal{A}_{k_{q-1}, k_q} f$ is, according to $\alpha$ and Theorem 1, well-defined and in $A_{\hat{k}}(\hat{S})$, for a sector $\hat{S}$ of finite radius, bisecting direction $d_q$, and opening more than $\pi/\kappa_q$, and $1/\hat{k} = 1/\tilde{k} + 1/\kappa_q$, and we require it to be admissible with respect to the parameter tupels $(k_1, \ldots, k_{q-1})$ and $(d_1, \ldots, d_{q-1})$ (meaning in particular, that $g$, which is defined for $\arg z = d_{q-1}$ and $|z|$ sufficiently small, *can be analytically continued* into a (small) sector of infinite radius and bisecting direction $d_{q-1}$).

We conclude directly from the definition that to every $f$ which is admissible with respect to $(k_1, \ldots, k_q)$ and $(d_1, \ldots, d_q)$, there correspond functions $f_0, \ldots, f_q$, iteratively defined by

$$
\begin{aligned}
f_q &= f \,, \\
f_j &= \mathcal{A}_{k_j, k_{j+1}} f_{j+1} \,, \quad 1 \le j \le q-1 \,, \\
f_0 &= \mathcal{L}_{k_1} f_1 \,.
\end{aligned}
$$

For $0 \le j \le q-1$ each $f_j$ is in $A_{\tilde{k}_j}(S_j)$, for a sector $S_j$ of finite radius, bisecting direction $d_{j+1}$, and opening larger than $\pi/\kappa_{j+1}$, with some $\tilde{k}_q > 0$ and

$$1/\tilde{k}_j = 1/\kappa_{j+1} + 1/\tilde{k}_{j+1} \,, \quad 1 \le j \le q-1 \,.$$

Moreover, each $f_j$ can be analytically continued into a (small) sector of infinite radius and bisecting direction $d_j$ and is of exponential size not larger than $\kappa_j$ there. Finally, if $\hat{f}_j = J(f_j)$ $(0 \leq j \leq q)$, then

$$\hat{f}_j = \hat{A}_{k_j, k_{j+1}} \hat{f}_{j+1}, \quad 1 \leq j \leq q-1,$$

$$\hat{f}_0 = \hat{\mathcal{L}}_{k_1} \hat{f}_1.$$

From now on, a vector $k = (k_1, \ldots, k_q)$ will be called *admissible*, if a) holds. A vector $d = (d_1, \ldots, d_q)$ will be named *admissible with respect to* $k$, if b) holds. If both applies, we say for short that $k$ and $d$ are admissible.

# Chapter 6

# Multisummable Power Series

We are now ready to give the definition of multisummable power series. Loosely speaking, $k$-summable power series are such that a formal application of the Borel operator of index $k$ produces a function to which the Laplace operator of the same index is applicable, and in multisummability, we replace the Laplace operator by a succession of (finitely many) acceleration operators. Doing so, we get a larger class of summable series, containing all formal solutions of non-linear ODE: this will be shown in Chapter 8.

## 6.1 Definition of Multisummability in Directions

Let a natural number $q$ and admissible parameter vectors $k = (k_1, \ldots, k_q)$, $d = (d_1, \ldots, d_q)$ be given. We call a formal series $\hat{f}$ $k$-summable in the multi-direction $d$, if $\hat{f} \in \mathbb{C}[[z]]_{1/k_q}$ (so that $\hat{B}_{k_q} \hat{f}$ is a convergent series), and if in addition the sum of $\hat{B}_{k_q} \hat{f}$ is admissible with respect to $k$ and $d$. The function

$$f = \mathcal{L}_{k_1} \circ \mathcal{A}_{k_1,k_2} \circ \ldots \circ \mathcal{A}_{k_{q-1},k_q} g,$$

$$g = \mathcal{S}(\hat{B}_{k_q} \hat{f})$$

(which is well-defined, according to Section 5.4) is called *the sum* of $\hat{f}$, and we write

$$f = \mathcal{S}_{k,d}(\hat{f}).$$

The set of all formal series being $k$-summable in the multi-direction $d$ will be denoted by

$$\mathbb{C}\{z\}_{k,d}.$$

**Remark.** It should be noted that for $q = 1$, the definitions given above coincide with the ones of Section 3.3. For $q \geq 2$, every $\hat{f} \in \mathbb{C}\{z\}_{k,d}$ gives rise to functions $f_0, \ldots, f_q$, with

$$f_q = \mathcal{S}(\hat{B}_{k_q} \hat{f}),$$

$$f_j = \mathcal{A}_{k_j,k_{j+1}} f_{j+1}, \quad 1 \leq j \leq q-1,$$

$$f_0 = \mathcal{L}_{k_1} f_1 \quad \left( = \mathcal{S}_{k,d}(\hat{f}) \right).$$

From Section 5.4 we conclude

$$f_j \in A_{\tilde{k}_j}(S_j), \quad 0 \le j \le q-1,$$

with $S_j$ a sector of finite radius, bisecting direction $d_{j+1}$ and opening larger than $\pi/\kappa_{j+1}$, and (with $\tilde{k}_q = \infty$)

$$1/\tilde{k}_j = 1/\kappa_{j+1} + 1/\tilde{k}_{j+1} = \sum_{\nu=j+1}^{q} 1/\kappa_\nu = 1/k_q - 1/k_j \; ;$$

in particular, $\tilde{k}_0 = k_q$. For $\hat{f}_j = J(f_j)$, $0 \le j \le q$, we find (using $\hat{B}_{k_j} = \hat{A}_{k_j, k_{j+1}} \circ \hat{B}_{k_{j+1}}$)

$$\hat{f}_j = \hat{B}_{k_j}(\hat{f}), \quad 1 \le j \le q,$$

so in particular, $\hat{f}_0 = \hat{f}$. Moreover, for $j = 1, \ldots, q$, each $f_j$ admits analytic continuation into a sector of infinite radius and bisecting direction $d_j$, and in this sector, $f_j$ is of exponential size not larger than $\kappa_j$.

In what follows, we call $\hat{f} \in \mathbb{C}[[z]]$ *multisummable in some multi-direction*, if admissible $k$ and $d$ exist so that $\hat{f} \in \mathbb{C}\{z\}_{k,d}$, and then we refer to $k$ as the *type* of multisummability.

## 6.2   Elementary Properties

The following properties of multisummability are direct consequences of the definition, and the proof can be left to the reader:

**Lemma 1.**   a)  *Let $\hat{f} \in \mathbb{C}\{z\}_{k,d}$ and $j$, $1 \le j \le q$, be given. Then there exists $\varepsilon > 0$ so that $f \in \mathbb{C}\{z\}_{k,\tilde{d}}$, for every $\tilde{d} = (d_1, \ldots, d_{j-1}, \tilde{d}_j, d_{j+1}, \ldots, d_q)$ satisfying*

$$|\tilde{d}_j - d_j| < \varepsilon$$

*and*

$$|\tilde{d}_j - d_{j-1}| \le \frac{\pi}{2\kappa_j}, \quad |d_{j+1} - \tilde{d}_j| \le \frac{\pi}{2\kappa_{j+1}} \; .$$

*Moreover,*

$$\mathcal{S}_{k,\tilde{d}}(\hat{f}) = \mathcal{S}_{k,d}(\hat{f}) \; .$$

b)  *Let $\hat{f} \in \mathbb{C}[[z]]$ be given, and let*

$$d = (d_1, \ldots, d_q), \quad \tilde{d} = (\tilde{d}_1, \ldots, \tilde{d}_q) \; ,$$

*be so that*

$$\tilde{d}_j = d_j + 2\pi, \quad 1 \le j \le n \; .$$

*Then (for every $k = (k_1, \ldots, k_q)$)*

$$\hat{f} \in \mathbb{C}\{z\}_{k,d} \quad \textit{iff} \quad \hat{f} \in \mathbb{C}\{z\}_{k,\tilde{d}},$$

*and if so, then*

$$S_{k,d}(\hat{f})(z) = S_{k,\tilde{d}}(\hat{f})(ze^{2\pi i}) ,$$

*for every z where either side is defined.*

Let admissible parameter vectors $k = (k_1, \ldots, k_q)$ and $d = (d_1, \ldots, d_q)$ (with $q \geq 2$) be given. Choose $\nu$, $1 \leq \nu \leq q$, and define

$$\tilde{k} = (k_1, \ldots, k_{\nu-1}, k_{\nu+1}, \ldots, k_q) ,$$
$$\tilde{d} = (d_1, \ldots, d_{\nu-1}, d_{\nu+1}, \ldots, d_q) .$$

Then it is easily seen that $\tilde{k}$ and $\tilde{d}$ are again admissible.

**Lemma 2.** *For $q \geq 2$ and $k,d$, resp. $\tilde{k}, \tilde{d}$ as above, we have*

$$\mathbb{C}\{z\}_{\tilde{k},\tilde{d}} \subset \mathbb{C}\{z\}_{k,d} ,$$

*and for every $\hat{f} \in \mathbb{C}\{z\}_{\tilde{k},\tilde{d}}$ and z with $|z|$ sufficiently small and $|d_1 - \arg z| \leq \pi/(2k_1)$, we have*

$$S_{\tilde{k},\tilde{d}}(\hat{f})(z) = S_{k,d}(\hat{f})(z) .$$

**Proof.** In case $\nu = q$, we find that $g = \mathcal{S}(\hat{\mathcal{B}}_{k_q}(\hat{f}))$ is entire and of exponential size at most $\kappa$, $1/\kappa = 1/k_{q-1} - 1/k_q$, in every sector of infinite radius. Hence we conclude from Ex. 5, Section 5.1, that $f = \mathcal{A}_{k_{q-1},k_q}(g) = \mathcal{S}(\hat{\mathcal{B}}_{k_{q-1}}(\hat{f}))$ (where integration in the acceleration operator may be along any ray, so in particular can be along $\arg z = d_q$). But this is sufficient to complete the proof in this case. In case $2 \leq \nu \leq q-1$, let

$$g = \mathcal{A}_{k_{\nu+1},k_{\nu+2}} \circ \ldots \circ \mathcal{A}_{k_{q-1},k_q} \circ \mathcal{S}(\hat{\mathcal{B}}_{k_q}(\hat{f}))$$

and $\tilde{g} = \mathcal{A}_{k_\nu,k_{\nu+1}}(g)$. Then from Ex. 2, Section 5.2, we find that $\mathcal{A}_{k_{\nu-1},k_\nu}(\tilde{g})$ exists (with integration along $\arg z = d_\nu$) and equals $\mathcal{A}_{k_{\nu-1},k_{\nu+1}}(g)$, and this is all we need to prove in this case. Finally, in case $\nu = 1$, we proceed as in the previous case, interpreting the (undefined) operators $\mathcal{A}_{k_0,k_1}$, resp. $\mathcal{A}_{k_0,k_2}$ as $\mathcal{L}_{k_1}$, resp. $\mathcal{L}_{k_2}$, and using Theorem 2, Section 5.2, instead of the exercise.                          □

**Exercises.** For admissible $k = (k_1, \ldots, k_q)$ and $d = (d_1, \ldots, d_q)$, with $q \geq 2$, we define *projections* $\pi_\nu$ by

$$\pi_\nu(k, d) = (\tilde{k}, \tilde{d}), \quad 1 \leq \nu \leq q ,$$

with $\tilde{k}, \tilde{d}$ as above.

1. For $1 \leq \nu_1 < \ldots < \nu_p \leq q$ $(1 \leq p \leq q-1)$, let

$$(\tilde{k}, \tilde{d}) = \pi_{\nu_1} \circ \ldots \circ \pi_{\nu_p}(k, d) .$$

Conclude from Lemma 2

$$\mathbb{C}\{z\}_{\tilde{k},\tilde{d}} \subset \mathbb{C}\{z\}_{k,d}$$

and interpret this for the special case of $p = q - 1$.

2. Let $(k, d)$ and $(\hat{k}, \hat{d})$ be two arbitrary admissible pairs of parameter vectors (not necessarily of the same dimension), and let

$$\hat{f}(z) \in \mathbb{C}\{z\}_{k,d} \cap \mathbb{C}\{z\}_{\hat{k},\hat{d}} \cdot$$

Moreover, assume that both $(k, d)$ and $(\hat{k}, \hat{d})$ may be obtained from the same admissible pair $(\tilde{k}, \tilde{d})$, applying suitable projections. Conclude from Lemma 2 that then

$$\mathcal{S}_{k,d}(\hat{f}) = \mathcal{S}_{\hat{k},\hat{d}}(\hat{f})$$

(in a suitable sector $S$).

3. For $k = (k_1, \ldots, k_q)$ and $d = (d_1, \ldots, d_q)$, $q \geq 2$, let

$$\hat{f} \in \mathbb{C}\{z\}_{k,d} \cap \mathbb{C}[[z]]_{1/k_{q-1}} \cdot$$

Show that then

$$\hat{f} \in \mathbb{C}\{z\}_{\tilde{k},\tilde{d}}, \quad (\tilde{k}, \tilde{d}) = \pi_q(k, d) \cdot$$

## 6.3    The Main Decomposition Result

It follows from Lemma 2 (see also Ex. 1 and 2, previous section) that for admissible $k = (k_1, \ldots, k_q)$ and $d = (d_1, \ldots, d_q)$, if $\hat{f}_j \in \mathbb{C}\{z\}_{k_j,d_j}$, $1 \leq j \leq q$, then

$$\hat{f} = \sum_{j=1}^{q} \hat{f}_j \in \mathbb{C}\{z\}_{k,d} \cdot$$

In this section we will show, under an additional assumption on $k$, that every $\hat{f} \in \mathbb{C}\{z\}_{k,d}$ is obtained in this fashion. To do so, we use the following

**Lemma 3.**    *Let real numbers $\tilde{k} > k > 0$ be given, so that $\tilde{k} > 1/2$, and define $\kappa$ by*

$$1/\kappa = 1/k - 1/\tilde{k} \cdot$$

*Let $\hat{f} \in \mathbb{C}\{z\}_{\tilde{k},d}$ (for some real number $d$), and let $\hat{g} = \hat{\mathcal{L}}_\kappa \hat{f}$. Then*

$$\hat{g} = \hat{g}_1 + \hat{g}_2 ,$$

*where $\hat{f}_1 := \hat{\mathcal{B}}_\kappa \hat{g}_1$ is convergent, and $\hat{g}_2 \in \mathbb{C}\{z\}_{k,d}$.*

**Proof.** By definition, $f = S_{\tilde{k},d}(\hat{f})$ is analytic in a sector $S = S(d,\alpha,r)$, $\alpha > \pi/\tilde{k}$, $r > 0$, and

$$f(z) \cong_{\tilde{k}} \hat{f}(z) \quad \text{in} \quad S \,.$$

For $\varepsilon > 0$, let $\tilde{S} = S(d,\tilde{\alpha},\tilde{r})$, $\tilde{\alpha} = \alpha - \varepsilon$, $\tilde{r} = r - \varepsilon$, and take $\varepsilon$ so small that $\tilde{\alpha} > \pi/\tilde{k}$ (and $\tilde{r} > 0$). Let $\gamma$ denote the positively oriented boundary of $\tilde{S}$, then by Cauchy's Integral Formula

$$f(z) = \frac{1}{2\pi i} \int_{\gamma} \frac{f(w)}{w-z} dw, \quad z \in \tilde{S} \,.$$

Decomposing $\gamma = \gamma_1 + \gamma_2$, where $\gamma_1$ is the circular part of $\gamma$, let

$$f_j(z) = \frac{1}{2\pi i} \int_{\gamma_j} \frac{f(w)}{w-z} dw, \quad z \in \tilde{S}, \quad j = 1,2 \,.$$

Then $f = f_1 + f_2$, and $f_1$ is analytic at the origin, and therefore

$$f_2(z) = f(z) - f_1(z) \cong_{\tilde{k}} \hat{f}_2(z) \quad \text{in} \quad \tilde{S} \,.$$

Moreover, $f_2$ remains analytic in $\hat{S} = S(d,\tilde{\alpha})$, tending to zero as $z \to \infty$ in $\hat{S}$. So Theorem 1, Section 2.1, implies

$$g_2(z) := (\mathcal{L}_\kappa f_2)(z) \cong_k \hat{g}_2(z) = (\hat{\mathcal{L}}_\kappa \hat{f}_2)(z)$$

in a sector of opening larger than $\pi/k$ and bisecting direction $d$. Hence by definition $\hat{g}_2 \in \mathbb{C}\{z\}_{k,d}$. Defining $\hat{g}_1 = \hat{g} - \hat{g}_2$, we have $\hat{f}_1 = \hat{B}_\kappa \hat{g}_1 = \hat{f} - \hat{f}_2$. Since $\hat{f}_1, \hat{f}_2 \in \mathbb{C}\{z\}_{\tilde{k},d}$, we find $\hat{f}_1 \in \mathbb{C}\{z\}_{\tilde{k},d}$, and $S_{\tilde{k},d}(\hat{f}_1) = f_1$. But $f_1$ is analytic at the origin, so $\hat{f}_1$ must converge. $\qquad\square$

**Theorem 1.** *Given admissible* $k = (k_1,\ldots,k_q)$ *and* $d = (d_1,\ldots,d_q)$, $q \geq 1$, *assume (with* $k_0 = +\infty$ *)*

$$1/k_j - 1/k_{j-1} < 2, \quad 1 \leq j \leq q \,.$$

*Then for* $\hat{f} \in \mathbb{C}\{z\}_{k,d}$ *we have*

$$\hat{f} = \sum_{j=1}^{q} \hat{f}_j \,,$$

*with* $\hat{f}_j \in \mathbb{C}\{z\}_{k_j,d_j}$, $1 \leq j \leq q$, *and*

$$S_{k,d}(\hat{f}) = \sum_{j=1}^{q} S_{k_j,d_j}(\hat{f}_j) \,.$$

**Proof.**   We proceed by induction with respect to $q$ : In case $q = 1$, the statement holds trivially, hence assume $q \geq 2$. Let

$$k = k_q, \quad 1/\tilde{k} = 1/k_q - 1/k_{q-1}$$

(hence $1/\kappa := 1/k - 1/\tilde{k} = 1/k_{q-1}$, i.e. $\kappa = k_{q-1}$) and observe that Theorem 1 of Section 5.2 implies

$$\mathcal{A}_{k_{q-1},k_q}(\mathcal{S}(\hat{\mathcal{B}}_{k_q}\hat{f}))(z) \cong_{\tilde{k}} \hat{\mathcal{B}}_{k_{q-1}}(\hat{f})(z) \quad \text{in} \quad S,$$

with a sector $S = S(d_q, \alpha, r)$ of opening $\alpha > \pi/k_{q-1}$. Therefore, $\hat{\mathcal{B}}_{k_{q-1}}(\hat{f}) \in \mathbb{C}\{z\}_{k_{q-1},d_q}$, hence Lemma 3 applies (with $d = d_q$) to $\hat{\mathcal{B}}_{k_{q-1}}(\hat{f})$ in place of $\hat{f}$, $\hat{f}$ in place of $\hat{g}$. So we find

$$\hat{f} = \hat{h} + \hat{f}_q,$$

with $\hat{f}_q \in \mathbb{C}\{z\}_{k_q,d_q}$, and $\hat{h}$ so that $\hat{\mathcal{B}}_{k_q-1}\hat{h}$ converges. Since $\hat{f}_q \in \mathbb{C}\{z\}_{k,d}$, we have

$$\hat{h} \in \mathbb{C}\{z\}_{k,d} \cap \mathbb{C}[[z]]_{1/k_{q-1}},$$

hence according to Ex. 3, previous section, we find

$$\hat{h} \in \mathbb{C}\{z\}_{\tilde{k},\tilde{d}}, \quad (\tilde{k},\tilde{d}) = \pi_q(k,d).$$

Applying the induction hypothesis to $h$, we see that the proof is completed.   □

**Exercises.**

  1. Let $k = (k_1, \ldots, k_q)$ and $d = (d_1, \ldots, d_q)$ be given, and define

$$\tilde{k} = pk = (pk_1, \ldots, pk_q), \quad \tilde{d} = d/p = (d_1/p, \ldots, d_q/p),$$

   for a natural number $p$.

   a) Show that $(k,d)$ is admissible iff $(\tilde{k},\tilde{d})$ is admissible.

   b) For admissible $(k,d)$, show

$$\hat{f}(z) \in \mathbb{C}\{z\}_{k,d} \quad \text{iff} \quad \hat{f}(z^p) \in \mathbb{C}\{z\}_{\tilde{k},\tilde{d}},$$

   and if so, then

$$\mathcal{S}_{\tilde{k},\tilde{d}}(\hat{f}(z^p)) = \mathcal{S}_{k,d}(\hat{f})(z^p).$$

In view of the previous exercise, we may extend the definition of multisummability to power series in a root $z^{1/p}$ $(p \geq 2)$ : A power series $\hat{f}$ in $z^{1/p}$ is called $k$-summable in the (multi-) direction $d$ iff $\hat{f}(z^p)$ is $pk$-summable in the (multi-) direction $d/p$.

2. Given admissible $k$ and $d$, and $\hat{f} \in \mathbb{C}\{z\}_{k,d}$, show for sufficiently large natural $p$ (depending only on $k$)

$$\hat{f} = \sum_{j=1}^{q} \hat{f}_j \, ,$$

with formal power series $\hat{f}_j$ in $z^{1/p}$ which are $k_j$-summable in direction $d_j$, $1 \le j \le q$.

3. For $k_1 > k_2 > 0$, $1/\kappa := 1/k_2 - 1/k_1 \ge 2$, let

$$\hat{f} = \hat{f}_1 + \hat{f}_2, \quad \hat{f}_j \in \mathbb{C}\{z\}_{k_j, d_j}, \quad j = 1, 2,$$

(with $|d_2 - d_1| \le \pi/(2\kappa)$, hence $\hat{f} \in \mathbb{C}\{z\}_{k,d}$, $k = (k_1, k_2)$, $d = (d_1, d_2)$). Show that then

$$\hat{g} = \hat{\mathcal{B}}_{k_1} \hat{f}$$

is $\kappa$-summable in direction $d_2$, hence

$$g = \mathcal{S}_{\kappa, d_2}(\hat{g})$$

is analytic in a sector $S = S(d_2, \alpha, r)$ of opening larger than $2\pi$. Moreover, show that

$$\psi(z) := g(z) - g(ze^{2\pi i})$$

(defined in $S(d, \alpha - 2\pi, r)$, $d = d_2 - \pi$) can be analytically continued into the sector (of infinite radius) $S(d, \alpha - 2\pi, +\infty)$.

4. For $k_1, k_2$ as in Ex. 3, show existence of

$$\hat{f} \in \mathbb{C}\{z\}_{k,d}, \quad k = (k_1, k_2), \quad d = (d_1, d_2)$$

which *cannot* be written as

$$\hat{f} = \hat{f}_1 + \hat{f}_2, \quad \hat{f}_j \in \mathbb{C}\{z\}_{k_j, d_j}, \quad j = 1, 2.$$

*Hint.* Show existence of $\psi(z)$, analytic in $S = S(d_2 - \pi, \alpha - 2\pi, r)$, having no analytic continuation beyond $|z| = r$, and so that $\psi(z) \cong_{\kappa} \hat{0}$ in $S$, and define $\hat{f}(z) = \sum f_n z^n$,

$$f_n = \frac{\Gamma(1 + n/k_1)}{2\pi i} \int_0^a \psi(w) w^{-n-1} dw, \quad n \ge 1$$

(for some $a \in S$).

5. Under the assumptions of Theorem 1, show that the decomposition of $\hat{f}$ into a sum $\sum_{j=1}^{q} \hat{f}_j$ is (for $q \geq 2$) never unique.

   *Hint.* Consider convergent series $\hat{g}_j$ with $\sum_{j=1}^{q} \hat{g}_j = \hat{0}$.

6. For $\hat{f} \in \mathbb{C}\{z\}_{k,d}$, $k = (k_1, \ldots, k_q)$, $d = (d_1, \ldots, d_q)$, and

$$\tilde{k} = (\tilde{k}_1, \ldots, \tilde{k}_{q-1}), \quad 1/\tilde{k}_j = 1/k_{j+1} - 1/k_1, \quad 1 \leq j \leq q-1,$$
$$\tilde{d} = (d_2, \ldots, d_q)$$

   show that

$$\hat{B}_{k_1}(\hat{f}) \in \mathbb{C}\{z\}_{\tilde{k},\tilde{d}},$$
$$\mathcal{L}_{k_1} \circ \mathcal{S}_{\tilde{k},\tilde{d}}(\hat{B}_k(\hat{f})) = \mathcal{S}_{k,d}(\hat{f}).$$

   *Hint.* Use Ex.1 to see that it suffices to consider the cases with $1/k_j - 1/k_{j-1} < 2$, and then use Theorem 1, together with Theorem 2 of Section 2.3.

## 6.4   The Algebra of Multisummable Power Series in a Direction

The following is the analogue to Theorem 2 in Section 3.3:

**Theorem 2.**   *For fixed, but arbitrary, admissible $k$ and $d$, we have:*

a) *If $\hat{f}$, $\hat{g} \in \mathbb{C}\{z\}_{k,d}$, then $\hat{f} + \hat{g}$, $\hat{f}\hat{g} \in \mathbb{C}\{z\}_{k,d}$, and*

$$\mathcal{S}_{k,d}(\hat{f} + \hat{g}) \quad = \quad \mathcal{S}_{k,d}(\hat{f}) + \mathcal{S}_{k,d}(\hat{g}),$$
$$\mathcal{S}_{k,d}(\hat{f}\hat{g}) \quad = \quad (\mathcal{S}_{k,d}(\hat{f}))(\mathcal{S}_{k,d}(\hat{g})).$$

b) *If $\hat{f} \in \mathbb{C}\{z\}_{k,d}$, then $\hat{f}'$, $\int_0^z \hat{f}(w)dw \in \mathbb{C}\{z\}_{k,d}$, and*

$$\mathcal{S}_{k,d}(\hat{f}') \quad = \quad \frac{d}{dz}\mathcal{S}_{k,d}(\hat{f}),$$
$$\mathcal{S}_{k,d}\left(\int_0^z \hat{f}(w)dw\right) \quad = \quad \int_0^z \mathcal{S}_{k,d}(\hat{f})(w)dw.$$

c) If $\hat{f} \in \mathbb{C}\{z\}_{k,d}$ has non-zero constant term, then $1/\hat{f} \in \mathbb{C}\{z\}_{k,d}$, and

$$\mathcal{S}_{k,d}(1/\hat{f}) = 1/(\mathcal{S}_{k,d}(\hat{f}))$$

(wherever the right hand side is defined).

d) If $\hat{f} \in \mathbb{C}\{z\}_{k,d}$ and $p$ is a natural number, then $\hat{f}(z^p) \in \mathbb{C}\{z\}_{pk,d/p}$, and

$$\mathcal{S}_{pk,d/p}(\hat{f}(z^p)) = \mathcal{S}_{k,d}(\hat{f})(z^p) .$$

**Proof.** Statement a) follows from the definition and (in case of multiplication) repeated application of Lemma 2 and Theorem 3 of Section 5.3. Statement d) is a consequence of Ex. 1, Section 6.3. To prove b), observe that the same exercise allows to assume without loss of generality

$$1/k_j - 1/k_{j-1} < 2, \quad 1 \leq j \leq q \quad (\text{with } k_0 = +\infty) .$$

In this situation, b) follows using Theorem 1 and the corresponding result on $k_j$-summability. Finally, c) follows from the definition of multisummability and Ex. 3, Section 5.3.

**Exercises.**

1. Show that the exercises in Section 3.3 generalize to multisummability.

2. Let $\hat{f}(z) = \sum\limits_{n=-p}^{\infty} f_n z^n$ be a formal Laurent series ($p$ a natural number) and let $k$ and $d$ be admissible. There are two ways of defining multisummability of $\hat{f}$ :

   a) We say that $\hat{f}$ is $k$-summable in the multi-direction $d$, if $\hat{g}(z) := \sum\limits_{n=0}^{\infty} f_n z^n \in \mathbb{C}\{z\}_{k,d}$, and we then define

   $$\mathcal{S}_{k,d}(\hat{f}) = \mathcal{S}_{k,d}(\hat{g}) + \sum\limits_{n=-p}^{-1} f_n z^n .$$

   b) We say that $\hat{f}$ is $k$-summable in the multi-direction $d$, if $\hat{g}(z) := z^p \hat{f}(z) \in \mathbb{C}\{z\}_{k,d}$, and we then define

   $$\mathcal{S}_{k,d}(\hat{f}) = z^{-p} \mathcal{S}_{k,d}(\hat{g}) .$$

   Show that both definitions are equivalent, and in case $f_{-p} = \ldots = f_{-1} = 0$ coincide with the original definition for power series.

## 6.5  Singular Directions

Let an admissible $k = (k_1, \ldots, k_q)$ and a formal power series $\hat{f}$ be given. A multi-direction $d$ (admissible with respect to $k$) will be called *singular* (for $\hat{f}$), iff $\hat{f} \notin \mathbb{C}\{z\}_{k,d}$; otherwise, we say that $d$ is *non-singular*. It is possible, even for $\hat{f} \in \mathbb{C}[[z]]_{1/k_q}$, that *all* $d$ are singular, e.g. when $g = S(\hat{\mathcal{B}}_{k_q}\hat{f})$ cannot be analytically continued across the boundary of some bounded region containing the origin. Inspecting the definition of $\mathbb{C}\{z\}_{k,d}$, one sees that the reason for $\hat{f} \notin \mathbb{C}\{z\}_{k,d}$ (i.e. $d$ singular) will be connected to the "level" $k_j$, leading to the following definition of singular directions of this level:

Given $k$ and $\hat{f}$ as above, a real number $d_j$ will be called *singular of level* $k_j$ (for fixed $j$, $1 \le j \le q$), iff we *can choose* $d_{j+1}, \ldots, d_q$ (in case of $j \le q-1$) with

$$|d_\nu - d_{\nu-1}| \le \frac{\pi}{2\kappa_\nu}, \quad j+1 \le \nu \le q$$

(see Section 5.4 for the definition of $\kappa_1, \ldots, \kappa_q$) so that the functions

$$f_j, \ldots, f_q$$

(as in Section 6.1) are well-defined, but so that $f_j$ *either cannot be analytically continued into a (small) sector of infinite radius and bisecting direction* $d_j$, *or in every such sector has exponential size larger than* $\kappa_j$.

It should be observed that, except for $j = q$, a value $d_j$ may be singular of level $k_j$, but for *appropriate* choices of $d_{j+1}, \ldots, d_q$ and $d_1, \ldots, d_{j-1}$, the multi-direction $d = (d_1, \ldots, d_q)$ is non-singular, while for some other choice of $d_{j+1}, \ldots, d_q$, *all* choices of $d_1, \ldots, d_{j-1}$ will lead to a singular multi-direction $d$. On the other hand, if $d = (d_1, \ldots, d_q)$ is admissible with respect to $k$, and all $d_j$ are non-singular of corresponding levels, then $d$ is a non-singular multi-direction.

Lemma 1 implies that $d_j$ is singular of level $k_j$ iff $d_j + 2\pi$ is, too, hence identifying values of $d_q \bmod 2\pi$, we say (as for $q = 1$) that $\hat{f}$ is $k$-*summable* iff it has at most finitely many singular directions of each level $k_j$, $1 \le j \le q$. The set of all $k$-summable $\hat{f}$ will be denoted by

$$\mathbb{C}\{z\}_k .$$

**Remark.** Let $\hat{f} \in \mathbb{C}\{z\}_k$, and let $d_j = \delta$ be singular of level $k_j$. For $j = q$, the reason for this can be that the function $g = f_q = S(\hat{\mathcal{B}}_{k_q}\hat{f})$ cannot be analytically continued along the ray $\arg z = \delta$; note that numbers close to $\delta$ are all non-singular, so if analytic continuation along $\arg z = \delta$ is possible, then $g$ is automatically analytic in a small sector of infinite radius bisected by $\arg z = \delta$. If $g(z)$ *can* be analytically continued along $\arg z = \delta$, then the only other reason for $\delta$ to be singular is that $g(z)$ is of infinite exponential size in every (small) sector bisected by this ray; note that along neighboring (hence non-singular) rays $g(z)$ is of exponential size at most $\kappa_q$, so if $g(z)$ would be of possibly large, but finite, exponential size in a sector bisected by $\arg z = \delta$, then Phragmen-Lindelöf's

Theorem ([CL], [Ti], [SG]) would imply the size to be at most $\kappa_q$, hence $\delta$ would be non-singular of level $k_j$.

For $j \leq q - 1$, choose non-singular directions $d_{j+1}, \ldots, d_q$ of respective levels $k_{j+1}, \ldots, k_q$ so that (with $d_j = \delta$)

$$|d_\nu - d_{\nu-1}| \leq \frac{\pi}{2\kappa_\nu}, \quad j+1 \leq \nu \leq q ;$$

such values exist according to the definition of singular directions of level $k_j$. Defining $g = f_j$ (with $f_j, \ldots, f_q$ as in Section 6.1) and arguing similarly as for the case $j = q$, one finds that for *at least one* choice of $d_{j+1}, \ldots, d_q$ as above, the function $g(z)$ either has a singularity on the ray $\arg z = \delta$, or is of infinite exponential size along this ray.

For $q = 1$, we have seen that absence of singular directions implies convergence of the series $\hat{f}$. This generalizes to arbitrary $q$ in the following way:

**Proposition 1.** *Let* $k = (k_1, \ldots, k_q)$, $q \geq 2$, *be admissible, and assume that* $\hat{f} \in \mathbb{C}\{z\}_k$ *has no singular directions of level* $k_j$, *for some fixed* $j$, $1 \leq j \leq q$. *Then* $\hat{f} \in \mathbb{C}\{z\}_{\tilde{k}}$, *with*

$$\tilde{k} = \pi_j(k) := (k_1, \ldots, k_{j-1}, k_{j+1}, \ldots, k_q) .$$

**Proof.** Consider the case $j = q$ : Absence of singular directions of this level implies that

$$g := \mathcal{S}(\hat{\mathcal{B}}_{k_q}(\hat{f}))$$

is entire and of exponential size $\kappa$, with

$$1/\kappa = 1/k_q - 1/k_{q-1} ,$$

in arbitrary sectors of infinite radius. Therefore, $\tilde{g} := \mathcal{A}_{k_{q-1},k_q} g$ can easily be seen to be analytic and single-valued in a punctured disc $D$ around the origin, and (see Theorem 1, Section 5.2)

$$\tilde{g}(z) \cong_{1/\kappa} (\hat{\mathcal{B}}_{k_{q-1}} \hat{f})(z) \quad \text{in} \quad D .$$

So $\tilde{g}$ is, in fact, analytic at the origin, and $\hat{\mathcal{B}}_{k_{q-1}} \hat{f}$ is its power series expansion. Hence $\hat{f} \in \mathbb{C}\{z\}_{\tilde{k}}$ follows from the definition.

In the other cases of $j \leq q - 1$, define

$$g := \mathcal{A}_{k_j,k_{j+1}} \circ \ldots \circ \mathcal{A}_{k_{q-1},k_q} (\mathcal{S}(\hat{\mathcal{B}}_{k_q}(\hat{f}))) ,$$

where integration is along arbitrarily fixed non-singular directions $d_{j+1}, \ldots, d_q$ (of corresponding levels $k_{j+1}, \ldots, k_q$). Then absence of singular directions of level $k_j$ implies that $g(z)$ can be analytically continued along all directions $\arg z = d_j$, with

$$|d_j - d_{j+1}| \leq \frac{\pi}{2\kappa_{j+1}}$$

and is of exponential size at most $\kappa_j$ in small sectors bisected by these rays. Hence we can define

$$\tilde{g} = \mathcal{A}_{k_{j-1},k_j} g$$

(integrating along $d_j$). Then Ex. 2, Section 5.2, implies (with $\mathcal{A}_{\infty,k_1} := \mathcal{L}_{k_1}$)

$$\tilde{g} = \mathcal{A}_{k_{j-1},k_{j+1}} \circ \mathcal{A}_{k_{j+1},k_{j+2}} \circ \ldots \circ \mathcal{A}_{k_{q-1},k_q}(\mathcal{S}(\hat{\mathcal{B}}_{k_q}\hat{f})) \, ,$$

and using this, we then find $\hat{f} \in \mathbb{C}\{z\}_{\tilde{k}}$. $\qquad\qquad\qquad\qquad\qquad\square$

**Exercises.**

1. Let $k = (k_1, \ldots, k_q)$, $q \geq 2$, be admissible, and let $\hat{f} \in \mathbb{C}\{z\}_k \cap \mathbb{C}[[z]]_{1/\kappa}$, for a real number $\kappa > k_q$. Show that $\hat{f}$ cannot have singular directions of level $k_q$ (hence $\hat{f} \in \mathbb{C}\{z\}_{\tilde{k}}$, $\tilde{k} = \pi_q(k)$).

2. Let $k = (k_1, \ldots, k_q)$ and $\tilde{k} = (\tilde{k}_1, \ldots, \tilde{k}_{\tilde{q}})$ both be admissible, with $\{k_1, \ldots, k_q\} \subset \{\tilde{k}_1, \ldots, \tilde{k}_{\tilde{q}}\}$. Show

$$\mathbb{C}\{z\}_k \subset \mathbb{C}\{z\}_{\tilde{k}} \, .$$

## 6.6 Multisummability

We will say that a formal power series $\hat{f}$ is *multisummable*, if we can find an admissible $k$ so that $\hat{f} \in \mathbb{C}\{z\}_k$. The following theorem shows that there always exists an optimal choice for $k$ (if $\hat{f}$ *is* multisummable).

**Theorem 3.** *Let* $k = (k_1, \ldots, k_q)$ *and* $\tilde{k} = (\tilde{k}_1, \ldots, \tilde{k}_{\tilde{q}})$ *both be admissible, and assume*

$$\hat{f} \in \mathbb{C}\{z\}_k \cap \mathbb{C}\{z\}_{\tilde{k}} \, .$$

*Then the following holds:*

a) *If* $\{k_1, \ldots, k_q\} \cap \{\tilde{k}_1, \ldots, \tilde{k}_{\tilde{q}}\} = \emptyset$, *then* $\hat{f}$ *converges.*

b) *In case* $\{k_1, \ldots, k_q\} \cap \{\tilde{k}_1, \ldots, \tilde{k}_{\tilde{q}}\} = \{\hat{k}_1, \ldots, \hat{k}_{\hat{q}}\} \neq \emptyset$, *assume without loss of generality*

$$\hat{k}_1 > \hat{k}_2 > \ldots > \hat{k}_{\hat{q}} \quad (> 0) \, ,$$

*so that* $\hat{k} = (\hat{k}_1, \ldots, \hat{k}_{\hat{q}})$ *is admissible. Then*

$$\hat{f} \in \mathbb{C}\{z\}_{\hat{k}} \, .$$

**Proof.** For $\hat{k} = k$, or $\hat{k} = \tilde{k}$, the statement $\hat{f} \in \mathbb{C}\{z\}_{\hat{k}}$ holds trivially, so we may restrict to cases where this is not so; hence we can find a smallest integer $j$,

$$0 \le j \le \min(q, \tilde{q}) - 1$$

for which $k_{q-j} \ne \tilde{k}_{\tilde{q}-j}$, and interchanging $k$ and $\tilde{k}$ if necessary, we may assume

$$k_{q-j} < \tilde{k}_{\tilde{q}-j} \ .$$

Assume $j = 0$. Then for $q \ge 2$, we may use the exercise at the end of the previous section to see

$$\hat{f} \in \mathbb{C}\{z\}_{k'} \cap \mathbb{C}\{z\}_{\tilde{k}}, \quad k' = \pi_q(k) \ ,$$

while for $q = 1$, Theorem 3 of Section 3.4 shows that $\hat{f}$ converges. Repeating this argument, one can prove a), while in case of b), we will either arrive at a situation where $j$ is positive, or at one of the trivial situations mentioned above.

Assume now $j \ge 1$. Then minimality of $j$ implies

$$k_{q-\nu} = \tilde{k}_{\tilde{q}-\nu}, \quad 0 \le \nu \le j - 1 \ .$$

Choosing an admissible $d = (d_1, \ldots, d_q)$ of non-singular directions (corresponding to $k$) of respective levels, we define

$$
\begin{aligned}
f_q &= S(\hat{B}_{k_q}\hat{f}) \ , \\
f_j &= A_{k_j, k_{j+1}} f_{j+1}, \quad 1 \le j \le q - 1 \ , \\
f_0 &= \mathcal{L}_{k_1} f_1 \ ,
\end{aligned}
$$

integrating along these directions. Then obviously,

$$
\begin{aligned}
f_q &= S(\hat{B}_{\tilde{k}_q}\hat{f}) \ , \\
f_{q-\nu} &= A_{\tilde{k}_{\tilde{q}-\nu}, \tilde{k}_{\tilde{q}-\nu+1}} f_{q-\nu+1}, \quad 1 \le \nu \le j \ ,
\end{aligned}
$$

so in particular, $d_{q-j+1}, \ldots, d_q$ are non-singular (corresponding to $\tilde{k}$) of respective levels. Altering $d_{q-j}$ slightly (if necessary), we may assume it non-singular of level $\tilde{k}_{\tilde{q}-j}$. This implies that $f_{q-\nu}$ can be analytically continued into a (small) sector $S = S(d_{q-j}, \alpha)$ and is of exponential size not more than $\tilde{\kappa}_{\tilde{q}-j}$, $1/\tilde{\kappa}_{\tilde{q}-j} = 1/\tilde{k}_{\tilde{q}-j} - 1/\tilde{k}_{\tilde{q}-j-1}$ (with $\tilde{k}_0 := +\infty$). This then shows that $f_{q-\nu-1} = A_{k_{q-\nu-1}, k_{q-\nu}} f_{q-\nu}$ is analytic in $\tilde{S} = S(d_{q-j}, \alpha + \pi/\kappa_{q-\nu})$, and is of *finite* exponential size there. The remark (in the previous section) on singular directions then shows that no singular directions of level $k_{q-\nu-1}$ can exist, so Proposition 1 implies

$$\hat{f} \in \mathbb{C}\{z\}_{k'} \cap \mathbb{C}\{z\}_{\tilde{k}}, \quad k' = \pi_{q-\nu-1}(k) \ .$$

Repeating this argument finitely many times, we will arrive at one of the trivial cases mentioned above, hence the proof is completed. □

**Remark 1.**  The definition for $\mathbb{C}\{z\}_k$ was here given in accordance with the concept of multi-directions $d = (d_1, \ldots, d_q)$ and a corresponding definition for singular directions. Instead, one can also consider multisummability *in a single direction*, meaning that we restrict to $d$ with $d_1 = \ldots = d_q$. This was done, e.g., in [Ba 1] and [MR] and leads to a different concept of singular directions, and a corresponding (different) definition for $\mathbb{C}\{z\}_k$. The analogon to Theorem 3, in this situation, which was announced without proof in [MR] and quoted (but not used) in [Ba 1], seems not to hold, although no counterexample is known as yet.

**Remark 2.**  Under the assumptions of Theorem 3, let $d = (d_1, \ldots, d_q)$, resp. $\tilde{d} = (\tilde{d}_1, \ldots, \tilde{d}_{\tilde{q}})$, be admissible with respect to $k$, resp. $\tilde{k}$, and so that

$$\hat{f} \in \mathbb{C}\{z\}_{k,d} \cap \mathbb{C}\{z\}_{\tilde{k},\tilde{d}} .$$

In case a), we then have

$$\mathcal{S}_{k,d}(\hat{f})(z) = \mathcal{S}_{\tilde{k},\tilde{d}}(\hat{f})(z) = \mathcal{S}(\hat{f})(z) ,$$

for every $z$ where the first two expressions are both defined. In case b), assume in addition that $k_j = \tilde{k}_\nu$ implies $d_j = \tilde{d}_\nu$, for all $j, \nu$ with $1 \le j \le q$, $1 \le \nu \le \tilde{q}$. Then Ex. 1, Section 6.2, implies

$$\mathcal{S}_{k,d}(\hat{f})(z) = \mathcal{S}_{\tilde{k},\tilde{d}}(\hat{f})(z) = \mathcal{S}_{\hat{k},\hat{d}}(\hat{f})(z)$$

(if $\hat{d}$ is defined so that $\hat{k}_j = k_\nu \; (= \tilde{k}_{\tilde{\nu}})$ implies $\hat{d}_j = d_\nu \; (= \tilde{d}_{\tilde{\nu}})$), for every $z$ where all three expressions are defined.

**Exercises.**

1. For $\hat{f}(z) = \sum\limits_{n=0}^{\infty} f_n z^n$ and $\varepsilon = e^{2\pi i/p}$, with integer $p \ge 2$, define

$$\hat{g}(z) = \hat{f}(z\varepsilon) = \sum_{n=0}^{\infty} f_n \varepsilon^n z^n .$$

Show $\hat{f} \in \mathbb{C}\{z\}_k$ implies $\hat{g} \in \mathbb{C}\{z\}_k$.

2. For $\hat{f}$ and $p$ as above, define

$$\hat{f}_j(z) = \sum_{n=0}^{\infty} f_{np+j} z^n , \quad j = 0, \ldots, p-1 .$$

Show $\hat{f} \in \mathbb{C}\{z\}_k$ iff $\hat{f}_j \in \mathbb{C}\{z\}_{k/p}$ for every $j = 0, \ldots, p-1$.

*Hint.* Use Ex. 1, Section 6.3.

3. For every $\hat{g} \in \mathbb{C}[[z]]$, show that there exists precisely one $\hat{f} \in \mathbb{C}[[z]]$, such that (formally)

$$z^3 \hat{f}'(z) + \hat{f}(z) = \hat{g}(z) .$$

4. With $\hat{f}$ and $\hat{g}$ as in Ex.3, let $\hat{f}_k(z) = \hat{\mathcal{B}}_k(\hat{f})(z)$, $\hat{g}_k(z) = \hat{\mathcal{B}}_k(\hat{g})(z)$, for $k = 1, 2$. Using Ex.1 and 2, Section 5.3, show that (formally)

$$\hat{f}_1(z) \quad = \quad \hat{g}_1(z) + \int_0^z (z - 2w)\hat{f}_1(w)dw ,$$

$$(2z^2 + 1)\hat{f}_2(z) \quad = \quad \hat{g}_2(z) + 4\int_0^z w\hat{f}_2(w)dw .$$

5. For $\hat{g}(z) = \sum_0^\infty (n!)z^n$, compute $g_1 = \mathcal{S}(\hat{\mathcal{B}}_1\hat{g})$ and show that the integral equation

$$f_1(z) = g_1(z) + \int_0^z (z - 2w)f_1(w)dw$$

has a unique solution $f_1$ which is analytic in the unit disc and can be analytically continued along every ray $\arg z = d_1$, $d_1 \neq 2k\pi$, $k \in \mathbb{Z}$. Moreover, show that $\hat{f}_1$ (defined as in Ex.4) is the power series expansion of $f_1$ (hence converges), and that $f_1$ is of exponential size not larger than two in (small) sectors bisected by the ray $\arg z = d_1$, $d_1$ as above.

*Hint.* Show that the integral equation is equivalent to the following initial value problem

$$f_1''(z) + zf_1'(z) = g_1''(z), \quad f_1(0) = g_1(0), \quad f_1'(0) = g_1'(0) ,$$

and derive the following representation for $f_1$ :

$$f_1(z) \quad = \quad g_1(z) + \int_0^z h(z, w)g_1(w)dw ,$$

$$h(z, w) \quad = \quad e^{w^2/2}\Big[\int_w^z e^{-u^2/2}du - w\Big] .$$

6. Conclude from Ex.5 that $g_2 = \mathcal{A}_{2,1}g_1$ and $f_2 = \mathcal{A}_{2,1}f_1$ (with integration along $\arg z = d_1$, $d_1$ as above) are in $\mathcal{A}_2(S)$, $S = S(d_1, \alpha, r)$, $\alpha > \pi/2$, $r > 0$ (finite), and satisfy the integral equation

$$(2z^2 + 1)f_2(z) = g_2(z) + 4\int_0^z wf_2(w)dw .$$

7. Show that $f_2(z)$ (as in Ex. 6) can be analytically continued along every ray $\arg z = d_2$, with $|d_1 - d_2| < \frac{\alpha}{2}$, $d_2 \neq \pi/2 + k\pi$, $k \in \mathbb{Z}$, and is of exponential size not larger than two in (small) sectors bisected by such rays.

*Hint.* As in Ex. 5, turn the integral equation into an initial value problem and compute its solution.

8. Conclude from the previous exercises that (for $\hat{g}$ as in Ex. 5, and $\hat{f}$ defined by Ex. 3)
$$\hat{f} \in \mathbb{C}\{z\}_k, \quad k = (2,1),$$
and find the singular rays of each level.

## 6.7  Some Applications of Cauchy-Heine Transforms

The following two Propositions characterize functions $f$ which are the sum of multisummable power series. To do so, it is worthwhile to define, for every admissible $k = (k_1, \ldots, k_q)$ and $d = (d_1, \ldots, d_q)$, closed intervals
$$I_j = [d_j - \pi/(2k_j), \ d_j + \pi/(2k_j)], \quad 1 \leq j \leq q,$$
and we say that $I_1, \ldots, I_q$ *correspond to* $d$. Admissibility of $d$ with respect to $k$ can then be seen to be equivalent to the inclusions
$$I_1 \subset I_2 \subset \ldots \subset I_q.$$
We also recall the convention $k_0 := +\infty$ and mention that
$$f(z) \cong_{k_0} \hat{f}(z) \quad \text{in} \quad S$$
implies $\hat{f}$ to converge and be the power series expansion of $f$ about the origin.

**Proposition 2.** *Let* $k = (k_1, \ldots, k_q)$ *and* $d = (d_1, \ldots, d_q)$ *be admissible. Then for every* $\hat{f} \in \mathbb{C}\{z\}_{k,d}$*, we can choose* $\varepsilon > 0$ *and* $r > 0$ *such that to every real* $\delta$ *there exists a function* $f(z; \delta)$*, analytic and asymptotic to* $\hat{f}(z)$ *of Gevrey order* $k_q$*, for*
$$z \in S_\delta := S(\delta, \varepsilon, r),$$
*such that for every* $\delta_1, \delta_2$ *with* $|\delta_1 - \delta_2| < \varepsilon$ *(i.e.* $S_{\delta_1} \cap S_{\delta_2} \neq \emptyset$*), and every* $j = 1, \ldots, q$ *we have: If* $\delta_1, \delta_2 \in I_j$*, then*
$$f(z; \delta_1) - f(z; \delta_2) \cong_{k_{j-1}} \hat{0} \quad \text{in} \quad S_{\delta_1} \cap S_{\delta_2}.$$
*In addition, if* $p$ *is a natural number with*
$$pk_q > 1/2,$$
*then we may choose the functions* $f(z; \delta)$ *so that for every* $\delta$
$$f(z; \delta) = f(ze^{2p\pi i}; \delta + 2p\pi) \quad \text{in} \quad S_\delta.$$

**Proof.**  Using Ex. 1 of Section 6.3, we see that we may, without loss of generality, restrict ourselves to cases with

$$k_q > 1/2$$

(so that for $p$ we may take 1). In this situation, Theorem 1 implies

$$\hat{f} = \sum_{j=1}^{q} \hat{f}_j , \quad \hat{f}_j \in \mathbb{C}\{z\}_{k_j, d_j} ,$$

hence

$$f(z) := \mathcal{S}_{k,d}(\hat{f})(z) = \sum_{j=1}^{q} f_j(z) ,$$

with $f_j = \mathcal{S}_{k_j, d_j}(\hat{f}_j)$ analytic and asymptotic to $\hat{f}_j$ of Gevrey order $k_j$ in $\tilde{S}_j = S(d_j, \alpha_j, r)$, $r > 0$, $\alpha_j > \pi/k_j$. Taking $\varepsilon > 0$ so small that $\delta \in I_j$ implies $S_\delta \subset \tilde{S}_j$, we define

$$f(z; \delta) = \sum_{j=1}^{q} f_j(z; \delta) ,$$

with

$$f_j(z; \delta) = f_j(z) \quad \text{if} \quad \delta \in I_j ,$$

and for $\delta \notin I_j$, we take *any* $f_j(z; \delta) \cong_{k_j} \hat{f}_j(z)$ in $S_\delta$. One can then easily check that all assertions of Proposition 2 hold (in particular, since $\delta$ and $\delta + 2\pi$ cannot both be in $I_j$, we can always arrange $f_j(z; \delta) = f_j(ze^{2\pi i}; \delta + 2\pi)$).  □

The next result is, in a sense, the converse of Proposition 2:

**Proposition 3.**  *Let* $k = (k_1, \ldots, k_q)$ *and* $d = (d_1, \ldots, d_q)$ *be admissible, and assume* $k_q > 1/2$. *For* $\varepsilon > 0$, $r > 0$ *and* $\delta_0$ *so that* $I_q \subset [\delta_0, \delta_0 + 2\pi]$, *assume existence of* $f(z; \delta)$ *(for every* $\delta$ *with* $\delta_0 \leq \delta \leq \delta_0 + 2\pi$*), analytic in* $S_\delta := S(\delta, \varepsilon, r)$ *and bounded at the origin, such that for every* $\delta_1, \delta_2$ *with* $|\delta_1 - \delta_2| < \varepsilon$ *(i.e.* $S_{\delta_1} \cap S_{\delta_2} \neq \emptyset$*) we have: If* $\delta_1, \delta_2 \in I_j$ *for some* $j$, $1 \leq j \leq q$, *then*

$$f(z; \delta_1) - f(z; \delta_2) \cong_{k_{j-1}} \hat{0} \quad \text{in} \quad S_{\delta_1} \cap S_{\delta_2} ;$$

*if either* $\delta_1$ *or* $\delta_2$ *is not in* $I_q$, *then*

$$f(z; \delta_1) - f(z; \delta_2) \cong_{k_q} \hat{0} \quad \text{in} \quad S_{\delta_1} \cap S_{\delta_2} .$$

*In addition, let*

$$f(z; \delta_0) = f(ze^{2\pi i}; \delta_0 + 2\pi) \quad \text{in} \quad S_{\delta_0} .$$

*Then there exists a (unique)* $\hat{f} \in \mathbb{C}\{z\}_{k,d}$ *with*

$$f(z; d_1) = \mathcal{S}_{k,d}(\hat{f})(z) \quad \text{in} \quad S_{d_1} .$$

*Moreover, if we define* $f(z; \delta)$ *for arbitrary real* $\delta$ *so that*

$$f(z, \delta) = f(ze^{2\pi i}; \delta + 2\pi) \quad \text{in} \quad S_\delta ,$$

*then* $f(z; \delta)$ *satisfy all the requirements of Proposition 2.*

**Proof.** With $\delta_1, \ldots, \delta_m$ so that

$$\delta_0 < \delta_1 < \ldots < \delta_{m-1} < \delta_m := \delta_0 + 2\pi \,,$$

$$\delta_j - \delta_{j-1} < \varepsilon, \quad 1 \le j \le m \,,$$

and that all the boundary points of all intervals $I_\nu$ occur in the partitioning, let $a_j = \rho \exp\{i(\delta_j + \delta_{j-1})/2\}$, $0 < \rho < r$, and define

$$g_j \;=\; CH_{a_j}(f(\cdot; \delta_j) - f(\cdot; \delta_{j-1})), \quad 1 \le j \le m \,,$$

$$f_j(z) \;=\; \sum_{\mu=1}^{j} g_\mu(z) + \sum_{\mu=j+1}^{m} g_\mu(ze^{2\pi i}), \quad 0 \le j \le m \,.$$

Then $g_j(z) \equiv 0$ whenever $\delta_j, \delta_{j-1} \in I_1$, and otherwise, if $\nu$ is taken minimally so that $\delta_j$ or $\delta_{j-1}$ is not in $I_\nu$ (note for later use that this implies $\delta_j \le d_\nu - \pi/(2k_\nu)$ or $\delta_{j-1} \ge d_\nu + \pi/(2k_\nu)$), then Proposition 1 of Section 4.1 implies

$$g_j(z) \;\widetilde{=}_{k_\nu}\; \hat{g}_j(z) \quad \text{in} \quad S(\tilde{d}_j, \tilde{\alpha}_j, \rho) \,,$$

with

$$\tilde{d}_j \;=\; \pi + (\delta_j + \delta_{j-1})/2 \,,$$

$$\tilde{\alpha}_j \;=\; 2\pi + \varepsilon - (\delta_j - \delta_{j-1}) \,,$$

$$\hat{g}_j \;=\; \widehat{CH}_{a_j}(f(\cdot; \delta_j) - f(\cdot; \delta_{j-1})), \quad 1 \le j \le m \,.$$

This implies

$$\hat{g}_j \in \mathbb{C}\{z\}_{k_\nu, d_\nu} \,,$$

and $S_{k_\nu, d_\nu}(\hat{g}_j)(z)$ equals $g_j(z)$, resp. $g_j(ze^{2\pi i})$, if $\delta_j \le d_\nu - \pi/(2k_\nu)$, resp. $\delta_{j-1} \ge d_\nu + \pi/(2k_\nu)$. Consequently,

$$\hat{g} := \sum_{j=1}^{m} \hat{g}_j \in \mathbb{C}\{z\}_{k,d} \,,$$

and

$$S_{k,d}(\hat{g})(z) = f_j(z) \,,$$

for every $j$ with $\delta_j \in I_1$. Moreover,

$$h(z) := f(z; \delta_j) - f_j(z)$$

can be seen to be independent of $j$, analytic and single-valued for $0 < |z| < \rho$, and bounded at the origin. So $\hat{h} = J(h)$ is convergent, hence

$$\hat{f} := \hat{h} + \hat{g} \in \mathbb{C}\{z\}_{k;d} \,,$$

and

$$S_{k,d}(\hat{f}) = f(\cdot; \delta_j) \,,$$

for every $j$ with $\delta_j \in I_1$, hence according to our assumptions

$$S_{k,d}(\hat{f}) = f(\cdot; \delta) \quad \text{for} \quad \delta \in I_1 \,,$$

and this completes the proof.                                                      $\square$

**Proposition 4.** *For admissible* $k = (k_1, \ldots, k_q)$ *with* $k_q > 1/2$, *let* $d^{(1)} = (d_1^{(1)}, \ldots, d_q^{(1)})$ *and* $d^{(2)} = (d_1^{(2)}, \ldots, d_q^{(2)})$ *both be admissible with respect to* $k$, *and so that*

$$|d_1^{(1)} - d_1^{(2)}| \leq \pi/k_1 .$$

*Then for* $\hat{f} \in \mathbb{C}\{z\}_{k,d^{(1)}} \cap \mathbb{C}\{z\}_{k,d^{(2)}}$, *the sums*

$$f^{(j)} = S_{k,d^{(j)}}(\hat{f}), \qquad j = 1, 2 ,$$

*are both defined in* $S = S(\delta, \alpha, r)$, *with* $r > 0$ *and*

$$\delta = (d_1^{(1)} + d_1^{(2)})/2, \quad \alpha > \pi/k_1 - |d_1^{(1)} - d_1^{(2)}| .$$

*For* $a \in S$, *we define*

$$\hat{g} = \widehat{CH}_a(f^{(1)} - f^{(2)}) ;$$

*then*

$$\hat{g} \in \mathbb{C}\{z\}_{k,d} ,$$

*for every* $d = (d_1, \ldots, d_q)$ *which is admissible with respect to* $k$ *and so that*

$$\max\{d_j^{(1)}, d_j^{(2)}\} \leq d_j \leq \min\{d_j^{(1)}, d_j^{(2)}\} + 2\pi, \quad 1 \leq j \leq q .$$

**Proof.** To simplify notation, let

$$d_j^+ = \max\{d_j^{(1)}, d_j^{(2)}\}, \quad d_j^- = \min\{d_j^{(1)}, d_j^{(2)}\}, \quad 1 \leq j \leq q .$$

For $\varepsilon > 0$ and $r > 0$ sufficiently small, and arbitrary real $\delta$, let $f^{(j)}(z; \delta)$ be as in Proposition 2 (with $d$ replaced by $d^{(j)}$), for $j = 1, 2$, and define

$$\psi(z; \delta) = f^{(1)}(z; \delta) - f^{(2)}(z; \delta) .$$

Then

$$\psi(z; \delta) \cong_{k_q} \hat{0} \quad \text{in} \quad S_\delta .$$

Let

$$I_j = \left[ d_j^+ - \frac{\pi}{2k_j}, \ d_j^- + \frac{\pi}{2k_j} \right], \quad 1 \leq j \leq q .$$

Then each $I_j$ is the intersection of the two intervals $I_j^{(1)}$ and $I_j^{(2)}$, corresponding to $d^{(1)}$ resp. $d^{(2)}$. Admissibility of $d^{(1)}, d^{(2)}$ with respect to $k$ and the fact that (by assumption) $I_1^{(1)}$ intersects with $I_1^{(2)}$ guarantee that

$$\emptyset \neq I_1 \subset I_2 \subset \ldots \subset I_q .$$

From Proposition 2, for $|\delta_1 - \delta_2| < \varepsilon$ and $j = 1, \ldots, q$, we find:
If $\delta_1, \delta_2 \in I_j$, then

$$\psi(z; \delta_1) - \psi(z; \delta_2) \cong_{k_{j-1}} \hat{0} \quad \text{in} \quad S_{\delta_1} \cap S_{\delta_2} .$$

For $\delta \in I_1$ we find (using Proposition 3) $\psi(z; \delta) = f^{(1)}(z) - f^{(2)}(z)$. This, together with Proposition 1 of Section 4.1, implies

$$g(z) := CH_a(f^{(1)} - f^{(2)})(z) \cong_{k_1} \hat{g}(z) \quad \text{in} \quad \tilde{S},$$

with $\tilde{S}$ of radius $\rho = |a|$ and bounded by the rays

$$\arg z = d_1^+ - \frac{\pi}{2k_1} - \frac{\varepsilon}{2} \quad \text{and} \quad \arg z = d_1^- + \frac{\pi}{2k_1} + \frac{\varepsilon}{2} + 2\pi .$$

This implies that $\hat{g}(z) \in \mathbb{C}\{z\}_{k_q, d_q}$, for every $d_q$ with

$$d_1^+ + \frac{\pi}{2}\left(\frac{1}{k_q} - \frac{1}{k_1}\right) \le d_q \le d_1^- + 2\pi - \frac{\pi}{2}\left(\frac{1}{k_q} - \frac{1}{k_1}\right),$$

hence $\hat{g} \in \mathbb{C}\{z\}_{k,d}$ for every admissible $d$ with $d_q$ as above.

It may happen that $d_1^- + \frac{\pi}{2k_1} = d_q^- + \frac{\pi}{2k_q}$, in which case one can omit the next arguments. In the other case of $d_1^- + \frac{\pi}{2k_1} < d_q^- + \frac{\pi}{2k_q}$, choose

$$\delta_0 = d_1^- + \frac{\pi}{2k_1} < \delta_1 < \ldots < \delta_m = d_q^- + \frac{\pi}{2k_q},$$

so that $\delta_j - \delta_{j-1} < \varepsilon$, $1 \le j \le m$, and that all the values

$$d_j^- + \frac{\pi}{2k_j}, \quad 2 \le j \le q-1,$$

occur among the values $\delta_1, \ldots, \delta_{m-1}$. For $a_j = \rho \exp\{i(\delta_j + \delta_{j-1})/2\}$, let

$$g_j \quad = \quad CH_{a_j}(\psi(\cdot; \delta_j)), \quad 0 \le j \le m,$$

$$g_{j,j-1} \quad = \quad CH_{a_j}(\psi(\cdot; \delta_j) - \psi(\cdot; \delta_{j-1})), \quad 1 \le j \le m.$$

Then one can see, for $\mu = 1, \ldots, m$, and

$$0 < |z| < \rho, \quad \delta_m - \varepsilon/2 < \arg z < \delta_0 + \varepsilon/2 + 2\pi,$$

that

$$g(z) = h_\mu(z) + \sum_{j=1}^{\mu} g_{j,j-1}(z) + g_\mu(z),$$

with $h_\mu(z)$ being analytic at the origin, hence

$$\hat{g}(z) = \hat{h}_\mu(z) + \sum_{j=1}^{\mu} \hat{g}_{j,j-1}(z) + \hat{g}_\mu(z),$$

where $\hat{h}_\mu$ is the (convergent) expansion of $h_\mu$, and $\hat{g}_\mu$, resp. $\hat{g}_{j,j-1}$ are the corresponding formal Cauchy-Heine Transforms. Arguing as above, one can see for $\nu = 1, \ldots, q-1$ and $j = 1, \ldots, \mu$:
If

$$d_\nu^- + \frac{\pi}{2k_\nu} \le \delta_{j-1} < \delta_j \le d_{\nu+1}^- + \frac{\pi}{2k_{\nu+1}},$$

then

$$g_{j,j-1}(z) \cong_{k_\nu} \hat{g}_{j,j-1}(z) \quad \text{in} \quad \tilde{S}_{j,j-1} \, ,$$

with $\tilde{S}_{j,j-1}$ bounded by $\arg z = \delta_j - \varepsilon/2$ and $\arg z = \delta_{j-1} + \varepsilon/2 + 2\pi$, hence $\hat{g}_{j,j-1} \in \mathbb{C}\{z\}_{k_\nu, d_\nu}$, for every $d_\nu$ with

$$d_{\nu+1}^- + \frac{\pi}{2}\Big(\frac{1}{k_{\nu+1}} + \frac{1}{k_\nu}\Big) \le d_\nu \le d_\nu^- + 2\pi \, ;$$

and moreover,

$$g_\mu(z) \cong_{k_q} \hat{g}_\mu(z) \quad \text{in} \quad \tilde{S}_\mu \, ,$$

with $\tilde{S}_\mu$ bounded by $\arg z = \delta_\mu - \varepsilon/2$ and $\arg z = \delta_\mu + \varepsilon/2 + 2\pi$, hence $\hat{g}_\mu(z) \in \mathbb{C}\{z\}_{k_q, d_q}$, for every $d_q$ with

$$\delta_\mu + \frac{\pi}{2k_q} \le d_q \le \delta_\mu + 2\pi - \frac{\pi}{2k_q} \, .$$

From above, we recall (using $\delta_0 = d_1^- + \pi/(2k_1)$) that $\hat{g} \in \mathbb{C}\{z\}_{k,d}$ for every admissible $d$ with

$$d_1^+ + \frac{\pi}{2}\Big(\frac{1}{k_q} - \frac{1}{k_1}\Big) \le d_q \le \delta_0 + 2\pi - \frac{\pi}{2k_q} \, ,$$

and proceeding inductively with respect to $\mu$, $(1 \le \mu \le m)$, we find the same for all admissible $d$ with

$$d_1^+ + \frac{\pi}{2}\Big(\frac{1}{k_q} - \frac{1}{k_1}\Big) \le d_q \le d_q^- + 2\pi \, ,$$

and the one sided conditions on the other $d_j$ :

$$d_j \le d_j^- + 2\pi, \quad 1 \le j \le q-1 \, .$$

Quite similar arguments can be used to extend the interval for $d_q$ to the left, and doing so, one finds the lower bounds for $d_j$, $1 \le j \le q-1$, thus completing the proof.      □

**Corollary to Proposition 4.** *For admissible $k = (k_1, \dots k_q)$ with $k_q > 1/2$, let $d^{(0)} = (d_1^{(0)}, \dots, d_q^{(0)})$ be admissible with respect to $k$. Moreover, for some $\varepsilon > 0$, let*

$$\hat{f} \in \mathbb{C}\{z\}_{k,d^+} \cap \mathbb{C}\{z\}_{k,d^-} \, ,$$

*for every $d^+$, resp. $d^-$, which is admissible with respect to $k$ and so that*

$$d_j^{(0)} < d_j^+ < d_j^{(0)} + \varepsilon, \quad 1 \le j \le q \, ,$$

*resp.*

$$d_j^{(0)} - \varepsilon < d_j^- < d_j^{(0)}, \quad 1 \le j \le q \, .$$

*With $f^+ = S_{k,d^+}(\hat{f})$, $f^- = S_{k,d^-}(\hat{f})$, we have that $f^\pm$ is independent of the particular choice of $d^+$, resp. $d^-$, and $\hat{g} = \widehat{CH}_a(f^+ - f^-)$ (with suitable $a$) is in $\mathbb{C}\{z\}_{k,d}$, for every $d$ with $d_j^{(0)} < d_j < d_j^{(0)} + 2\pi$.*

**Remark.** We wish to point out that, in case $q = 1$, the above Corollary implies $\hat{g} \in \mathbb{C}\{z\}_{k_1}$, with at most one singular direction, namely $d_1^{(0)}$. This is not so for $q > 1$ : *There may exist admissible multi-directions $d$ with, say $d_j < d_j^{(0)}$, but $d_{j+1} > d_{j+1}^{(0)}$, for some $j$, $1 \le j \le q-1$, and nothing is said about $k$-summability of $\hat{g}$ in this multi-direction.*

**Exercises.**

1. In addition to the assumptions of the above Corollary, suppose $f^+ \equiv f^-$. Show that then
$$\hat{f} \in \mathbb{C}\{z\}_{k,d^{(0)}} \, .$$
   *Hint.* Use the exercises in Section 5.2.

2. Under the assumptions of the above Corollary, show $\hat{f} - \hat{g} \in \mathbb{C}\{z\}_{k,d}$, for every $d$, admissible with respect to $k$ and so that
$$|d_j^{(0)} - d_j| < \varepsilon, \quad 1 \le j \le q \, ,$$
   hence in particular for $d = d^{(0)}$.

3. For $\psi(z) = (1 - z)^{-1} \exp\{-z^{-2}\}$, show that $\hat{f}(z) = \sum_0^\infty f_n z^n$,
$$f_n = \frac{\Gamma(1 + n/2)}{2\pi i} \int\limits_0^{1/2} \psi(w) w^{-n-1} dw \, ,$$
   is in $\mathbb{C}\{z\}_k$, $k = (2,1)$.

4. Let $\hat{f} = \hat{f}_1 + \hat{f}_2$, $\hat{f}_j \in \mathbb{C}\{z\}_{k_j}$, $j = 1,2$ (with $k_1 > k_2 > 0$), and let $g = S(\hat{\mathcal{B}}_{k_2}(\hat{f}))$. Let $\delta$ be singular of level $k_2$, and define $h_1$, resp. $h_2$ by application of the acceleration operator $\mathcal{A}_{k_1,k_2}$ to $g$, with integration along $\arg z = d$, and $d$ slightly larger, resp. smaller, than $\delta$. Show that
$$\psi = h_1 - h_2$$
   is analytic in the sector $S = S(\delta, \pi/\kappa)$, $1/\kappa = 1/k_2 - 1/k_1$.

5. Conclude from Ex. 3 and 4 that $\hat{f} \in \mathbb{C}\{z\}_k$, $k = (2,1)$, exists, which is *not* the sum of $\hat{f}_1 \in \mathbb{C}\{z\}_2$ and $\hat{f}_2 \in \mathbb{C}\{z\}_1$.

# Chapter 7

# Some Equivalent Definitions of Multisummability

The following equivalent formulations of multisummability are, in some situations, more easily used in proving multisummability of certain series, or in (numerical) computations of the sum. To prove equivalence, we will heavily rely upon the main decomposition result of Section 6.3.

## 7.1   An Inductive Definition

Based upon Ex. 6 in Section 6.3, we can give the following inductive definition of multisummability:

**Proposition 1.**   *Given admissible* $k = (k_1, \ldots, k_q)$ *and* $d = (d_1, \ldots, d_q)$, *with* $q \geq 2$, *the following two assertions are equivalent:*

i) *The formal power series* $\hat{f}$ *is in* $\mathbb{C}\{z\}_{k,d}$.

ii) *For* $\tilde{k} = (\tilde{k}_1, \ldots, \tilde{k}_{q-1})$, $1/\tilde{k}_j = 1/k_{j+1} - 1/k_1$, $1 \leq j \leq q-1$, *and* $\tilde{d} = (d_2, \ldots, d_q)$, *we have* $\hat{g} := \hat{B}_{k_1}(\hat{f}) \in \mathbb{C}\{z\}_{\tilde{k},\tilde{d}}$. *Moreover,* $g := \mathcal{S}_{\tilde{k},\tilde{d}}\hat{g}$ *can be analytically continued into a (small) sector of infinite radius and bisecting direction* $d_1$, *and there,* $g$ *is of exponential size not more than* $k_1$.

*In case* i) *or* ii) *hold, we have*

$$\mathcal{S}_{k,d}(\hat{f}) = \mathcal{L}_{k_1} \circ \mathcal{S}_{\tilde{k},\tilde{d}}(\hat{g}) .$$

**Proof.**   Ex. 1 of Section 6.3 shows that we may, without loss of generality, assume $1/k_j - 1/k_{j-1} < 2$, $1 \leq j \leq q$ (with $k_0 = +\infty$). In this case, if i) holds, then Theorem 1 of Section 6.3 implies

$$\hat{f} = \sum_{j=1}^{q} \hat{f}_j , \quad \hat{f}_j \in \mathbb{C}\{z\}_{k_j, d_j} , \quad 1 \leq j \leq q ,$$

and

$$f := \mathcal{S}_{k,d}(\hat{f}) = \sum_{j=1}^{q} f_j , \quad f_j = \mathcal{S}_{k_j,d_j}(\hat{f}_j), \quad 1 \le j \le q .$$

From Theorem 2 of Section 2.3 we then conclude

$$g := \mathcal{B}_{k_1}(f) = \sum_{j=1}^{q} g_j , \quad g_j = \mathcal{B}_{k_1}(f_j), \quad 1 \le j \le q ,$$

where $g_1$ is analytic at the origin, and

$$g_j = \mathcal{S}_{\tilde{k}_{j-1},d_j}(\hat{g}_j), \quad \hat{g}_j = \hat{\mathcal{B}}_{k_1}(\hat{f}_j), \quad 2 \le j \le q .$$

This implies ii). Conversely, if ii) holds, then Theorem 1 of Section 6.3 implies (with $\hat{g}_j$ different from above)

$$\hat{g} = \sum_{j=1}^{q-1} \hat{g}_j , \quad \hat{g}_j \in \mathbb{C}\{z\}_{\tilde{k}_j,d_{j+1}} , \quad 1 \le j \le q-1 .$$

From Lemma 3 in Section 6.3 we find

$$\hat{\mathcal{L}}_{k_1}\hat{g}_j = \hat{f}_{j+1} + \hat{\mathcal{L}}_{k_1}\hat{h}_j , \quad 1 \le j \le q-1 ,$$

with $\hat{f}_{j+1} \in \mathbb{C}\{z\}_{k_{j+1},d_{j+1}}$, and $\hat{h}_j$ convergent. This implies

$$\hat{g} = \hat{h} + \sum_{j=2}^{q} \hat{\mathcal{B}}_{k_1}(\hat{f}_j) ,$$

with $\hat{h} = \sum_{j=1}^{q-1} \hat{h}_j$ convergent, and $\hat{\mathcal{B}}_{k_1}(\hat{f}_j) \in \mathbb{C}\{z\}_{\tilde{k}_{j-1},d_j}$. Therefore,

$$g = \mathcal{S}_{\tilde{k},\tilde{d}}(\hat{g}) = \mathcal{S}(\hat{h}) + \tilde{g} ,$$

$$\tilde{g} = \sum_{j=2}^{q} \mathcal{S}_{\tilde{k}_{j-1},d_j}(\hat{\mathcal{B}}_{k_1}\hat{f}_j) .$$

So we conclude that $\mathcal{S}(\hat{h})$ can be analytically continued into the same sector as $g$ and is also of exponential size at most $k_1$. This then shows that $\hat{f}_1 = \hat{\mathcal{L}}_{k_1}\hat{h} \in \mathbb{C}\{z\}_{k_1,d_1}$, hence

$$\hat{f} = \sum_{j=1}^{q} \hat{f}_j \in \mathbb{C}\{z\}_{k,d} ,$$

and

$$\mathcal{S}_{k,d}(\hat{f}) = \mathcal{L}_{k_1}(\mathcal{S}_{\tilde{k}_1,\tilde{d}}\hat{g}) . \qquad \square$$

**Exercises.**

1. Use Theorem 1 to give a simple proof that $\hat{f}$, as in Ex. 3 of Section 6.7, is in $\mathbb{C}\{z\}_k$, $k = (2,1)$.

2. Let $\hat{f} \in \mathbb{C}\{z\}_{k,d}$, for admissible $k$ and $d$. For $\kappa > 0$, let $\hat{g}_\kappa := \hat{\mathcal{B}}_\kappa(\hat{f})$. Show that $\hat{g}_\kappa$ is

   a) convergent for $\kappa \le k_q$,

   b) in $\mathbb{C}\{z\}_{\tilde{k},\tilde{d}}$, with

$$\left. \begin{array}{rcl} 1/\tilde{k}_j & = & 1/k_{j+\nu} - 1/\kappa \\ \tilde{d}_j & = & d_{j+\nu} \end{array} \right\} \; 1 \le j \le q - \nu \,,$$

   for $k_\nu \ge \kappa > k_{\nu+1}$, with some $\nu$, $0 \le \nu \le q-1$ (taking $k_0 = +\infty$).

# 7.2 Summability through Iterated Laplace Transforms

In theory, the use of Ecalle's acceleration operators in the definition of multisummability is no problem, and the original definition is particularly well-suited for proving summability of the product of two multisummable power series. However, in (numerical) computations of the sum of a multisummable formal series, it will be better to use the following, equivalent, method, which avoids acceleration operators, and instead uses an iteration of Laplace Transforms.

**Theorem 1.** *Given admissible* $k = (k_1, \ldots, k_q)$ *and* $d = (d_1, \ldots, d_q)$, *and defining* $\kappa_1, \ldots, \kappa_q$ *as in Section 5.4, the following two statements are equivalent:*

i) *The formal power series* $\hat{f}$ *is in* $\mathbb{C}\{z\}_{k,d}$.

ii) *The series* $\hat{g}_q := \hat{\mathcal{B}}_{\kappa_1} \circ \ldots \circ \hat{\mathcal{B}}_{\kappa_q} \hat{f}$ *converges, and* $g_q := S(\hat{g}_q)$ *is analytic and of exponential size* $\kappa_q$ *in a (small) sector of infinite radius and bisecting direction* $d_q$ *(so that* $\mathcal{L}_{\kappa_q} g_q$ *is defined). Moreover, for* $j = q$, *resp.* $q-1, \ldots$, *resp.* 2, *the function*

$$g_{j-1} = \mathcal{L}_{\kappa_j} g_j$$

*is analytic and of exponential size* $\kappa_{j-1}$ *in a (small) sector of infinite radius and bisecting direction* $d_{j-1}$ *(so that* $\mathcal{L}_{\kappa_{j-1}} g_{j-1}$ *is defined).*

*In case* i) *or* ii) *hold, we have*

$$S_{k,d}(\hat{f}) = \mathcal{L}_{\kappa_1} \circ \ldots \circ \mathcal{L}_{\kappa_q} g_q \,,$$

*integrating in directions* $d_1, \ldots, d_q$.

**Proof.** We proceed by induction with respect to $q$ : For $q = 1$, nothing is to be shown. For $q \geq 2$, we use Proposition 1 and the induction hypothesis for $\tilde{k}, \tilde{d}$ to see that i) and ii) are equivalent (also observe $\kappa_1 = k_1$). This then completes the proof.

**Exercises.** A sequence $(\lambda_n)_{n=0}^{\infty}$ will be called a *summability factor* for $\mathbb{C}\{z\}_{k,d}$ iff

$$\hat{f} = \sum_0^{\infty} f_n z^n \in \mathbb{C}\{z\}_{k,d}$$

implies

$$\hat{g}_\lambda := \sum_0^{\infty} f_n \lambda_n z^n \in \mathbb{C}\{z\}_{k,d} .$$

1. For admissible $k$ and $d$ (and $\kappa_1, \ldots, \kappa_q$ as in Section 5.4), show that for fixed $j$, $1 \leq j \leq q$, the sequence

$$\left( \frac{\Gamma(1 + n/\kappa_j)\Gamma(1 + n/k_{j-1})}{\Gamma(1 + n/k_j)} \right)_{n=0}^{\infty}$$

is a summability factor for $\mathbb{C}\{z\}_{k,d}$.

2. For $k, d$ and $\kappa_1, \ldots, \kappa_q$ as above, show that for fixed $j$, $1 \leq j \leq q$, the sequence

$$\left( \frac{\Gamma(1 + n/k_j)}{\Gamma(1 + n/\kappa_j)\Gamma(1 + n/k_{j-1})} \right)_{n=0}^{\infty}$$

is a summability factor for $\mathbb{C}\{z\}_{k,d}$.

3. For $k$ and $d$ as above, and $\alpha, \beta > 0$, show that

$$\left( \frac{\Gamma(1 + \alpha n)\Gamma(1 + \beta n)}{\Gamma(1 + (\alpha + \beta)n)} \right)_{n=0}^{\infty},$$

as well as

$$\left( \frac{\Gamma(1 + (\alpha + \beta)n)}{\Gamma(1 + \alpha n)\Gamma(1 + \beta n)} \right)_{n=0}^{\infty}$$

are summability factors for $\mathbb{C}\{z\}_{k,d}$.

*Hint.* Take admissible $\tilde{k}, \tilde{d}$ so that

$$\alpha^{-1}, (\alpha + \beta)^{-1}, k_1, \ldots, k_q \in \{\tilde{k}_1, \ldots, \tilde{k}_{\tilde{q}}\} ,$$

and apply Ex. 1,2 to $\mathbb{C}\{z\}_{\tilde{k},\tilde{d}}$; then use Ex. 2 to come back to $\mathbb{C}\{z\}_{k,d}$.

## 7.3 A Definition via the Sum

The following is just a reformulation of Propositions 2 and 3 of Section 6.7:

**Theorem 2.** *Given admissible* $k = (k_1, \ldots, k_q)$ *and* $d = (d_1, \ldots, d_q)$, *and assuming* $k > 1/2$, *the following statements are equivalent (with* $I_j$ *as in Section 6.7):*

i) *The formal power series* $\hat{f}$ *is in* $\mathbb{C}\{z\}_{k,d}$.

ii) *For* $\hat{f} \in \mathbb{C}[[z]]$ *we can choose* $\varepsilon > 0$ *and* $r > 0$ *such that to every real* $\delta$ *there exists a function* $f(z; \delta)$, *analytic and asymptotic to* $\hat{f}$ *of Gevrey order* $k_q$, *for* $z \in S_\delta := S(\delta, \varepsilon, r)$, *such that for every* $\delta_1, \delta_2$ *with* $|\delta_1 - \delta_2| < \varepsilon$ *and every* $j = 1, \ldots, q$, $\delta_1, \delta_2 \in I_j$ *implies*

$$f(z; \delta_1) - f(z; \delta_2) \underset{k_{j-1}}{\cong} \hat{0} \quad in \quad S_{\delta_1} \cap S_{\delta_2} .$$

iii) *For* $\hat{f} \in \mathbb{C}[[z]]$ *we can choose* $\varepsilon > 0$, $r > 0$ *and* $\delta_0$ *with* $I_q \subset [\delta_0, \delta_0 + 2\pi]$ *such that to every* $\delta$ *with*

$$\delta_0 \leq \delta \leq \delta_0 + 2\pi$$

*there exists a function* $f(z; \delta)$, *analytic in* $S_\delta := S(\delta, \varepsilon, r)$ *and bounded at the origin, such that for every* $\delta_1, \delta_2$ *with* $|\delta_1 - \delta_2| < \varepsilon$ *and every* $j = 1, \ldots, q$, $\delta_1, \delta_2 \in I_j$ *implies*

$$f(z; \delta_1) - f(z; \delta_2) \underset{k_{j-1}}{\cong} \hat{0} \quad in \quad S_{\delta_1} \cap S_{\delta_2} ,$$

*while otherwise*

$$f(z; \delta_1) - f(z; \delta_2) \underset{k_q}{\cong} \hat{0} \quad in \quad S_{\delta_1} \cap S_{\delta_2} .$$

*Additionally, we have*

$$f(z; \delta_0) = f(ze^{2\pi i}; \delta_0 + 2\pi) \quad in \quad S_{\delta_0} ,$$

*and*

$$f(z; d_1) \underset{k_q}{\cong} \hat{f}(z) \quad in \quad S_{d_1} .$$

**Remark 1.** Equivalence of i) and ii), and the following Theorem 3, are contained in [MgR].

As an application of Theorem 2, we prove

**Theorem 3.**   *For natural $n$, assume $h(z, y_1, \ldots, y_n)$ to be analytic in every variable, for*

$$|z| < \rho, \quad |y_j| < \rho, \quad j = 1, \ldots, n$$

*(with suitably small $\rho > 0$). Expanding $h$ into its power series in the variables $z, y_1, \ldots, y_n$, and choosing arbitrary formal power series $\hat{f}_1, \ldots, \hat{f}_n$ with vanishing constant terms, one can (formally) define a power series*

$$\hat{g}(z) = h(z, \hat{f}_1(z), \ldots, \hat{f}_n(z)) \, .$$

*For arbitrary admissible $k$ and $d$ with $k_q > 1/2$, if*

$$\hat{f}_1, \ldots, \hat{f}_n \in \mathbb{C}\{z\}_{k,d} \, ,$$

*then*

$$\hat{g} \in \mathbb{C}\{z\}_{k,d}$$

*follows.*

**Proof.**   Since $\hat{f}_j$ has vanishing constant term, the series $\hat{f}_j^\nu$, for natural $\nu$, starts with terms $z^\nu$ or later. Hence replacing $y_j^{\nu_j}$ by $\hat{f}_j^{\nu_j}$ (in the power series expansion of $g$), one obtains a finite sum of terms corresponding to every fixed power of $z$. Therefore, $\hat{g}$ is well-defined. To $\hat{f}_1, \ldots, \hat{f}_n \in \mathbb{C}\{z\}_{k,d}$, we can, according to Theorem 2, find $\varepsilon > 0$, $r > 0$ and $\delta_0$ (with $I_q \subset [\delta_0, \; \delta_0 + 2\pi]$) such that to every $\delta$ with $\delta_0 \leq \delta \leq \delta_0 + 2\pi$ and every $\nu$, $1 \leq \nu \leq m$, there exists a function $f_j(z; \delta)$, analytic in $S_\delta$ and so that $z f_\nu(z; \delta)$ is bounded at the origin (in $S_\delta$). Moreover, for every $\delta_1, \delta_2$ with $|\delta_1 - \delta_2| < \varepsilon$, and every $j = 1, \ldots, q$, $\delta_1, \delta_2 \in I_j$ implies

$$f_\nu(z; \delta_1) - f_\nu(z; \delta_2) \, \widetilde{=}_{k_{j-1}} \, \hat{0} \quad \text{in} \quad S_{\delta_1} \cap S_{\delta_2} \, ,$$

while otherwise

$$f_\nu(z; \delta_1) - f_\nu(z; \delta_2) \, \widetilde{=}_{k_q} \, \hat{0} \quad \text{in} \quad S_{\delta_1} \cap S_{\delta_2} \, .$$

Moreover, we have

$$f_\nu(z; \delta_0) = f_\nu(z e^{2\pi i}; \delta_0 + 2\pi) \quad \text{in} \quad S_{\delta_0} \, ,$$

and

$$f_\nu(z; d_1) \, \widetilde{=}_{k_q} \, \hat{f}_\nu(z) \quad \text{in} \quad S_{d_1} \, .$$

Defining

$$g(z; \delta) = h(z, f_1(z; \delta), \ldots, f_n(z; \delta)) \, ,$$

we find that $g(z; \delta)$ is analytic in $\tilde{S}_\delta = S(\delta, \varepsilon, \tilde{r})$, for suitably small $\tilde{r} > 0$, and

$$g(z; \delta_0) = g(z e^{2\pi i}; \delta_0 + 2\pi) \, .$$

Since $h(z, y_1, \ldots, y_n)$ satisfies a Lipschitz condition with respect to $(y_1, \ldots, y_n)$, one finds for $\delta_1, \delta_2$ and $j$ as above that $\delta_1, \delta_2 \in I_j$ implies

$$g(z; \delta_1) - g(z; \delta_2) \, \widetilde{=}_{k_{j-1}} \, \hat{0} \quad \text{in} \quad S_{\delta_1} \cap S_{\delta_2} \, ,$$

while otherwise

$$g(z; \delta_1) - g(z; \delta_2) \,\widetilde{=}_{k_q}\, \hat{0} \quad \text{in} \quad S_{\delta_1} \cap S_{\delta_2} \,.$$

This, in view of Proposition 3 of Section 6.7, implies the existence of *some* $\hat{g}_1 \in \mathbb{C}\{z\}_{k,d}$, with

$$g(z; d_1) = S_{k,d}(\hat{g}_1) \,,$$

hence in particular,

$$g(z; d_1) \,\widetilde{=}\, \hat{g}_1(z) \quad \text{in} \quad S_{d_1} \,.$$

But this can easily be seen to imply $\hat{g}_1 = \hat{g}$. □

**Exercises.**

1. Let $r$ be a natural number and $a \in \mathbb{C}\backslash\{0\}$. Let $k = (k_1, \ldots, k_q)$ and $d = (d_1, \ldots, d_q)$ be admissible, and so that

   a) for some (unique) $j_0$, $1 \le j_0 \le q$, we have $r = k_{j_0}$,

   b) $\arg(-a) \not\equiv r d_{j_0} \mod 2\pi$.

   For $\hat{f} \in \mathbb{C}\{z\}_{k,d}$, define $\hat{g}$ by

   $$(z^{r+1}\frac{d}{dz} + a)\hat{g} = \hat{f} \,.$$

   Show $\hat{g} \in \mathbb{C}\{z\}_{k,d}$.

2. For complex $a$ with Re $a > 0$, and arbitrary admissible $k$ and $d$, let $\hat{f} \in \mathbb{C}\{z\}_{k,d}$ and define $\hat{g}$ by

   $$(z\frac{d}{dz} + a)\hat{g} = \hat{f} \,.$$

   Show $\hat{g} \in \mathbb{C}\{z\}_{k,d}$.

# Chapter 8

# Formal Solutions to Non-Linear ODE

As an application of the theory of multisummability, we will now prove that formal solutions to non-linear meromorphic ODE are always multisummable. The first complete proof of this fact was given by Braaksma [Br 1]. A different proof, based on cohomological arguments, is due to Ramis and Sibuya [RS 2].

The proof presented here is as follows: We first normalize the differential equation to obtain a situation where there is *exactly one* formal solution. This normalization follows the same lines as in Braaksma's proof, but the remaining arguments are essentially different: Roughly speaking, for normalized equations we employ an iteration process, similar to Picard-Lindelöf's proof of existence and uniqueness of solutions to boundary value problems. Unfortunately, the situation here is more complicated in that the iteration does not produce *one* sequence of functions in *one* region, but we have to simultaneously consider finitely many regions, and in each region define a sequence of functions; moreover, we have to keep track of how these functions are interrelated. An additional complication is created by the fact that the regions mentioned above cannot be taken to be sectors, because the estimates used to give uniform convergence of the corresponding sequences seem to fail for sectors — at least, we have not been able to show otherwise.

## 8.1 Normalizations of the Equation

Throughout this chapter, we are concerned with systems of non-linear meromorphic ordinary differential equations

$$z^{r+1}y' = g(z,y),$$

where $r$ is a fixed natural number (the *Poincaré rank* of the equation), and $g(z,y) = g(z,y_1,\ldots,y_n)$ is an $n$-vector of functions

$$g_j(z,y) = g_j(z,y_1,\ldots,y_n), \quad 1 \le j \le n,$$

which are assumed analytic (and single-valued) in the polydisc

$$0 \le |z| < \rho, \quad 0 \le |y_j| < \rho, \quad 1 \le j \le n$$

(for some $\rho > 0$). *Assuming existence* of a formal power series solution

$$\hat{y}(z) = \sum_{m=1}^{\infty} y_m z^m, \quad y_m \in \mathbb{C}^n, \ m \geq 1,$$

(with vanishing constant term), we will show that every such solution is multi-summable. To identify the type of summability and possible singular rays, we have to suitably normalize the equation:

With $M$ and $\mu$ to be chosen later, we substitute

$$y = p(z) + z^\mu \tilde{y}, \quad p(z) = \sum_{m=1}^{M+\mu-1} y_m z^m$$

into the equation and obtain

$$\begin{aligned}
z^{r+1}\tilde{y}\,' &= \tilde{g}(z, \tilde{y}), \\
\tilde{g}(z, \tilde{y}) &= z^{-\mu}\{g(z, y) - \mu z^{\mu+r}\tilde{y} - z^{r+1}p'(z)\} \\
&= \tilde{g}_0(z) + (\tilde{A}(z) - \mu z^r)\tilde{y} + \tilde{g}_2(z, \tilde{y})
\end{aligned}$$

where

$$\begin{aligned}
\tilde{g}_0(z) &= \tilde{g}(z, 0) = z^{-\mu}\{g(z, p(z)) - z^{r+1}p'(z)\}, \\
\tilde{A}(z) &= \left.\frac{\partial g}{\partial y}\right|_{y=p(z)},
\end{aligned}$$

and $\tilde{g}_2(z, \tilde{y})$ accordingly. It follows that $z^{-\mu}\tilde{g}_2(z, \tilde{y})$ is analytic (in all variables, particularly in $z$) at the origin, and

$$z^{-\mu}\tilde{g}_2(z, \tilde{y}) = O(\|\tilde{y}\|_\infty^2) \quad \text{as} \quad \tilde{y} \to 0$$

(the $O$-constant being locally uniform in $z$). Moreover, $\tilde{A}(z)$ is analytic at the origin, and if we take $M + \mu$ sufficiently large, then an arbitrarily prescribed, but finite, number of terms in the expansion of $\tilde{A}(z)$ about the origin is independent of $M$ and $\mu$. Finally, since the new equation has the formal solution

$$\sum_{m=M}^{\infty} y_{m+\mu} z^m,$$

we conclude that $z^{-M}\tilde{g}_0(z)$ is analytic at the origin.

The, by now classical, formal theory of linear systems (which may, to the degree required here, be found, e.g., in [Wa]), assures that, after a change of variable $z \mapsto z^p$, with suitable integer $p$, which we assume made before we started, we can find a matrix polynomial

$$T(z) = \sum_{k=0}^{N} T_k z^k,$$

with

$$\det T(z) = \sum_{k=\nu_0}^{Nn} t_k z^k , \quad t_{\nu_0} \neq 0 ,$$

for some $0 \leq \nu_0 \leq Nn$, so that the transformed matrix

$$B(z) := T^{-1}(z)[\tilde{A}(z)T(z) - z^{r+1}T'(z)]$$

has the following form:

$$B(z) = \text{diag}\,[b_1(z), \ldots, b_n(z)] + z^r C(z) ,$$

where

$$b_j(z) = \sum_{\nu=r_j}^{r-1} b_{j,\nu} z^\nu , \quad b_{j,r_j} \neq 0, \ 1 \leq j \leq n ,$$

with

$$r \geq r_1 \geq r_2 \geq \ldots \geq r_n \geq 0$$

(for $r_j = r$, this should be read as $b_j(z) \equiv 0$), and $C(z)$ is analytic at the origin, with $C(0)$ commuting with $\text{diag}\,[b_1(z), \ldots, b_n(z)]$. Since $T(z)$ depends only upon finitely many terms in the expansion of $\tilde{A}(z)$ about the origin, we find that $T(z)$, in particular $\nu_0$, and then $b_1(z), \ldots, b_n(z)$ and $C(0)$, *are independent of $\mu$ and $N$*, provided $\mu + N$ is sufficiently large. Hence substituting

$$\tilde{y} = T(z)x ,$$

we find

$$
\begin{aligned}
z^{r+1}x' &= f(z, x) , \\
f(z, x) &= T^{-1}(z)\{\tilde{g}(z, T(z)x) - z^{r+1}T'(z)x\} \\
&= f_0(z) + (B(z) - \mu z^r)x + f_2(z, x) , \\
f_0(z) &= T^{-1}(z)\tilde{g}_0(z) , \\
f_2(z, x) &= T^{-1}(z)\tilde{g}_2(z, T(z)x) .
\end{aligned}
$$

Obviously, $z^{\nu_0}T^{-1}(z)$ is analytic at the origin (and $\nu_0$ is independent of $M$ and $\mu$, as we remarked before), hence we have that $z^{\nu_0-M}f_0(z)$ and $z^{\nu_0-\mu}f_2(z, x)$ are analytic (in $z$) in a full neighborhood of the origin. By taking $M$ and $\mu$ large enough, we see that the same applies to $z^{-r-1}f_0(z)$ and $z^{-r-1}f_2(z, x)$. This shows that the system $z^{r+1}x' = f(z, x)$ can be written as follows:

Let $k_1 > k_2 > \ldots > k_q > 0 =: k_{q+1}$ denote the *different* ones among the integers $r_1, \ldots, r_n$, and block the vectors $x$ and $f$ into blocks, so that for each entry in a block the corresponding $r_j$ coincide (if no value $r_j$ equals zero, we will formally introduce a block with number $q+1$ of dimension zero). Then the system looks as follows:

$$(z^K \delta + A)x = z(b(z) + g(z,x)) ,$$

where

$$K = \text{diag} \, [k_1 I_{s_1}, \ldots, k_{q+1} I_{s_{q+1}}] ,$$

with integers $s_j \geq 1$, $1 \leq j \leq q$, and $s_{q+1} \geq 0$,

$$A = \text{diag} \, [A_1, \ldots, A_{q+1}] ,$$

with $A_j$ being constant invertible matrices of dimension $s_j$, and for $j = q+1$ (if $s_{q+1} > 0$), all eigenvalues of $A_{q+1}$ have positive real parts, and $b(z)$ and $g(z,x)$ (different from the $g(z,y)$ considered above) are analytic in all variables in the region

$$|z| < \rho_0 , \qquad \|x\|_\infty < \rho_0 ,$$

with suitable $\rho_0 > 0$, and

$$g(z,0) \equiv 0 .$$

This system has the formal solution

$$\hat{x}(z) \quad = \quad T^{-1}(z) \sum_{m=M}^{\infty} y_{m+\mu} z^m$$

$$= \quad \sum_{m=M-\nu_0}^{\infty} x_m z^m , \quad x_m \in \mathbb{C}^n ,$$

and once we have proven $\hat{x}(z)$ to be multisummable, then so is the original formal solution $\hat{y}(z)$, and type of multisummability as well as singular directions of $\hat{x}$ and $\hat{y}$ obviously are the same. So it suffices to consider *normalized systems*, i.e. systems of the above special form.

**Remark 1.**   While the original system might have several different formal power series solutions (this happens even for linear systems), a normalized system can only have *one*, as can be seen by inserting a formal series and comparing coefficients.

**Remark 2.**   In view of the exercise below, we can, and will, assume in the sequel that at least one of the numbers $r_j$ was positive, i.e. that $q \geq 1$. We emphasize again that it may happen that *all* $r_j$ are positive, in which case the dimension of the matrix $A_{q+1}$ has to be taken zero.

**Remark 3.**   It can be seen from above that we may restrict to normalized systems having a formal solution

$$\hat{x}(z) = \sum_{m=m_0}^{\infty} x_m z^m ,$$

with $m_0$ being arbitrarily large. Inserting $\hat{x}(z)$ into the system and comparing coefficients, one can see that this is equivalent to

$$b_j(z) = O(z^{m_0-1}) \quad \text{as} \quad z \to 0 ,$$

for $j = 1, \ldots, q+1$. In what follows, we will only use $m_0 \geq 1$.

**Exercise.**

Consider a normalized system of ODE in case $q = 0$ (i.e. all entries $r_j = 0$). Show that then its (unique) formal power series solution $\hat{x}(z)$ converges.

*Hint.* Show existence and uniqueness of a sequence $x(z; \nu)$, analytic for $|z| < \rho_1$, with suitably small $\rho_1 \le \rho_0$, and so that $x(z; 0) \equiv 0$, and for $\nu \ge 0$

$$(z\frac{d}{dz} + A)x(z; \nu + 1) = z(b(z) + g(z, x(z; \nu)))\,.$$

For $|z| \le \rho_1$, show uniform convergence to an analytic solution of the system, and conclude that $\hat{x}(z)$ is its power series expansion.

## 8.2   Formal Iterations

Let a normalized system of ODE (see the previous section) be arbitrarily given, and assume $q \ge 1$. Beginning with

$$\hat{x}(z; 0) = \hat{0}\,,$$

we define a sequence $\hat{x}(z; \nu)$, $\nu \ge 0$, of formal (vector) power series by the identities

$$(z^K \delta + A)\hat{x}(z; \nu + 1) = z(b(z) + g(z, \hat{x}(z; \nu)))\,, \quad \nu \ge 0\,.$$

It follows easily that this determines $\hat{x}(z; \nu)$ uniquely, and the constant terms all vanish. Investigating the differences

$$\hat{x}(z; \nu + 1) - \hat{x}(z; \nu)\,,$$

one can show by induction that for $\nu \ge m$, the coefficients of $z^m$ in $\hat{x}(z; \nu)$ do no longer depend upon $\nu$, hence in this manner we can *prove existence of a formal solution to every normalized system.* This, however, is not our main concern, but instead, we will show next (by induction with respect to $\nu$) that

$$\hat{x}(z; \nu) \in \mathbb{C}\{z\}_k^n\,, \quad \nu \ge 0\,.$$

by which we mean to say that each component of the vector $\hat{x}(z; \nu)$ is in $\mathbb{C}\{z\}_k$. In view of Theorem 3 in Section 7.3, this follows from

**Lemma 1.** *Let $r$ be a non-negative integer, and let $A$ be an invertible $s \times s$ matrix, whose eigenvalues, in case $r = 0$, have positive real parts. Let $k = (k_1, \ldots, k_q)$ and $d = (d_1, \ldots, d_q)$ be admissible, and so that in case $r > 0$ we have*

$$r \in \{k_1, \ldots, k_q\}\,,$$

*and no eigenvalues of $-A$ are on the ray $\arg z = rd_j$, with $j$ so that $k_j = r$. For $\hat{f} \in \mathbb{C}[[z]]^s$, define $\hat{g} \in \mathbb{C}[[z]]^s$ by*

$$(z^{r+1}\frac{d}{dz} + A)\hat{g} = \hat{f}\,.$$

*Then $\hat{f} \in \mathbb{C}\{z\}_{k,d}^s$ implies $\hat{g} \in \mathbb{C}\{z\}_{k,d}^s$.*

**Proof.** For an invertible constant matrix $T$ we have $\hat{g} \in \mathbb{C}\{z\}_{k,d}^s$ iff $T\hat{g} \in \mathbb{C}\{z\}_{k,d}^s$. Using this, we find that it suffices to consider matrices $A$ which are in Jordan form; in fact, we may assume that $A$ is just one Jordan block. Writing out the resulting identities for the elements of the vector $\hat{g}$, we see that it suffices to treat the one-dimensional case, and in this situation, the proof follows from the exercises in Section 7.3. □

In what follows, we wish to show that the sums of the series $\hat{x}(z;\nu)$ converge uniformly on suitable regions. To do so, and to describe these regions, it is more convenient to study our normalized differential equation in a vicinity of $\infty$ rather than zero: Replacing $z$ by $1/z$, we obtain a normalized equation of the form

$$(A - z^{-K}\delta)x = z^{-1}(b(z) + g(z,x)) \, ,$$

with functions $b(z)$ and $g(z,x)$ (different from the ones above), which are analytic in all variables, for

$$1/\rho_0 < |z| \leq \infty, \quad 0 \leq \|x\|_\infty < \rho_0 \, ,$$

with suitably small $\rho_0 > 0$, and $g(z,0) \equiv 0$, $|z| > 1/\rho_0$. Consequently, a sector $S = S(d,\alpha,\rho)$ will from now on be the set of $z$ with

$$|z| > \rho^{-1}, \quad |d + \arg z| < \alpha/2 \, .$$

**Exercises.** In what follows, let $r$ be natural, $a = |a|e^{i\varphi} \neq 0$, and $\rho$ and $\varepsilon$ positive (real). Let

$$G \subset \{z \mid |z| > \rho^{-1}\}$$

be a region (on the Riemann surface of the Logarithm), so that for every $z \in G$ a real number $\tau$ with

$$|r\tau + \varphi| \leq \pi/2 - \varepsilon$$

exists, such that the curve $\gamma_z$ with representation

$$w(x) = [z^r + (xe^{i\tau})^r]^{1/r}, \quad x \geq 0 \, ,$$

is completely contained in $G$. Moreover, the set of all such $\tau$, corresponding to a given $z \in G$, is required to be connected (i.e. an interval). Every such region $G$ will be called *admissible* (with respect to $r, a, \rho$, and $\varepsilon$), and every such $\tau$ will be called *an admissible direction* (with respect to $G, \varepsilon$, and $z \in G$).

1. For $r = 1$, picture admissible regions $G$. Show that a maximal admissible region $G$ exists, having the following form:

   In the sector $S(\varphi, 3\pi - \varepsilon, \rho)$, draw halflines parallel to the two boundary rays and tangent to the circle $|z| = \rho^{-1}$. Let $G$ be the region bounded by part of the circle and these two halflines.

2. For $r \geq 2$, show that $G$ is admissible with respect to $r, a, \rho$ and $\varepsilon$ iff

$$\tilde{G} = \{z \mid z^r \in G\}$$

is admissible with respect to $1, a, \rho^{1/r}$ and $\varepsilon$. Use this to find maximal admissible regions corresponding to $r, a, \rho$ and $\varepsilon$.

3. Let $G$ be an admissible region, and assume $f$ to be analytic in $G$, and

$$\|f\|_\infty := \sup_{z \in G} |f(z)| < \infty .$$

Define

$$g(z) = \exp\{\frac{a}{r} z^r\} \int_z^{\infty(\tau)} \exp\{-\frac{a}{r} w^r\} f(w) w^{r-1} dw ,$$

integrating along a curve as above, for some admissible $\tau$. Show that $g$ is analytic in $G$ and independent of $\tau$. Moreover, show existence of a constant $C > 0$, depending upon $a, \varepsilon, r$, but independent of $\rho, \tau$ and $f$, so that

$$\|g\|_\infty \leq C \|f\|_\infty .$$

4. Let $G$ be an admissible region, and assume

$$G = \bigcup_{\nu=1}^m G_\nu , \quad m \geq 2 ,$$

with regions $G_\nu$ so that for every $z \in G_\nu$ and every (real) $x \geq 1$ we have $xz \in G_\nu$ $(1 \leq \nu \leq m)$. For $\nu = 1, \ldots, m$, let $f_\nu$ be analytic and bounded in $G_\nu$, and assume existence of constants $k > r$ and $c > 0$ so that for every $\nu$ and $\mu$ with $1 \leq \nu < \mu \leq m$

$$\exp\{c|z|^k\}(f_\nu(z) - f_\mu(z))$$

remain bounded in $G_\nu \cap G_\mu$ (which holds trivially if $G_\nu \cap G_\mu = \emptyset$). Moreover, assume that $c$ and $a$ (as above) are such that

$$\exp\{az^r\} \cong 0 \quad \text{as} \quad z \to \infty \quad \text{in} \quad G_\nu \cap G_\mu$$

implies

$$\exp\{az^r\} \leq \exp\{-c|z|^r\} \quad \text{in} \quad G_\nu \cap G_\mu ,$$

for $1 \leq \nu < \mu \leq m$.

(a) Show existence and uniqueness of functions $g_\nu$, $\nu = 1, \ldots, m$, which have the same properties as the $f_\nu$ (with the same constants $k$ and $c$) and satisfy (for $1 \leq \nu \leq m$)

$$(a - z^{-r}\delta)g_\nu(z) = f_\nu(z) \quad \text{in} \quad G_\nu .$$

*Hint.* Assuming existence, show for $1 \leq \nu < \mu \leq m$

$$g_\nu(z) - g_\mu(z) = \exp\{\frac{a}{r}z^r\} \int_z^\infty \exp\{-\frac{a}{r}w^r\}(f_\nu(w) - f_\mu(w))w^{r-1}dw ,$$

in $G_\nu \cap G_\mu$, and for $1 \leq \nu \leq m$ (with arbitrary $z_\nu \in G_\nu$)

$$g_\nu(z) = \exp\{\frac{a}{r}z^r\}[c_\nu + \int_{z_\nu}^z \exp\{-\frac{a}{r}w^r\}f_\nu(w)w^{r-1}dw]$$

in $G_\nu$, with a suitable constant $c_\nu$. From these identities, conclude uniqueness of $g_1, \ldots, g_m$. To show existence, let $z \in G_\nu$ and $\gamma = \gamma(z)$ be *any* path from $z$ to $\infty$ for which $\exp\{-\frac{a}{r}w^r\} \xrightarrow{\sim} 0$ as $w \to \infty$ on $\gamma(z)$ (such paths exist because of admissibility of $G$). Decompose $\gamma = \gamma_0 + \gamma_1 + \ldots + \gamma_\ell$, with $\gamma_\mu$ completely contained in one region $G_{\nu(\mu)}$. For $1 \leq \mu \leq \ell$, let $z_\mu$ be the initial point of $\gamma_\mu$, and define

$$g_\nu(z)\exp\{-\frac{a}{r}z^r\} = \sum_{\mu=0}^\ell \int_{\gamma_\mu} \exp\{-\frac{a}{r}w^r\}f_{\nu(\mu)}(w)w^{r-1}dw$$

$$+ \sum_{\mu=1}^\ell \int_{z_\mu}^\infty \exp\{-\frac{a}{r}w^r\}(f_{\nu(\mu)}(w) - f_{\nu(\mu-1)}(w))w^{r-1}dw ,$$

where in the second group of integrals we integrate along $\arg w = \arg z_\mu$. Show that this definition does not depend upon the decomposition of $\gamma$ (nor upon the choice of regions $G_{\nu(\mu)}$ to cover the individual pieces). Having done so, it is easy to obtain independence of $g_\nu$ of the choice of $\gamma(z)$ and then to derive the desired properties.

(b) Defining

$$\|(f_1, \ldots, f_m)\| \quad := \quad \sup_{1 \leq \nu \leq m} \sup_{z \in G_\nu} |f_\nu(z)|$$

$$+ \quad \sup_{1 \leq \nu < \mu \leq m} \sup_{z \in G_\nu \cap G_\mu} \exp\{c|z|^k\}|f_\nu(z) - f_\mu(z)|$$

(ignoring terms with $G_\nu \cap G_\mu = \emptyset$), show existence of $C > 0$, depending upon $a, \varepsilon, r$, but independent of $\rho$ and $f_1, \ldots, f_m$, so that

$$\|(g_1, \ldots, g_m)\| \leq C\|(f_1, \ldots, f_m)\| .$$

*Hint.* First estimate $g_\nu(z) - g_\mu(z)$, then choose $\gamma(z)$ to be as in the definition of admissibility, and write the representation for $g_\nu(z)$ in the form

$$g_\nu(z)\exp\{-\frac{a}{r}z^r\} \;=\; \sum_{\mu=0}^{\ell}\int_{\gamma_\mu}\exp\{-\frac{a}{r}w^r\}f_{\nu(\mu)}(w)w^{r-1}dw$$

$$+ \sum_{\mu=1}^{\ell}\exp\{-\frac{a}{r}z_\mu^r\}(g_{\nu(\mu)}(z_\mu) - g_{\nu(\mu-1)}(z_\mu))\,;$$

then estimate this formula, observing in particular that all the points $z_\mu$ lie on the curve $\gamma(z)$, implying

$$|\exp\{\frac{a}{r}(z^r - z_\mu^r)\}| \le 1\,.$$

(c) Suppose $\tilde{G}_\nu$ $(1 \le \nu \le m)$ are so that

$$z \in G_\nu \quad \text{is equivalent to} \quad ze^{2\pi i} \in \tilde{G}_\nu\,,$$

and let $\tilde{f}_\nu$ satisfy

$$\tilde{f}_\nu(ze^{2\pi i}) = f_\nu(z), \quad z \in G_\nu\,.$$

Define $\tilde{g}_\nu$ (on $\tilde{G}_\nu$) as in (a), and show that then

$$\tilde{g}_\nu(ze^{2\pi i}) = g_\nu(z), \quad z \in G_\nu\,.$$

## 8.3 Definition of Admissible Regions

Let an equation

$$(A - z^{-K}\delta)x = z^{-1}(b(z) + g(z,x))$$

be given, with $A = \text{diag}[A_1,\ldots,A_{q+1}]$ and $K = \text{diag}[k_1 I_{s_1},\ldots,k_{q+1}I_{s_{q+1}}]$ as described in Section 8.1, and $b(z)$, $g(z,x)$ analytic for

$$\rho_0^{-1} < |z| \le \infty, \quad \|x\|_\infty < \rho_0\,.$$

To prove multisummability of the formal solution $\hat{x}(z)$ (which now is a power series in $z^{-1}$, with vanishing constant term), we will, roughly speaking, define a sequence $x(z;\nu)$, $\nu \ge 0$, of functions, analytic and bounded in a region $G$ (independent of $\nu$), so that $x(z;0) \equiv 0$ and

$$(A - z^{-K}\delta)x(z;\nu+1) = z^{-1}(b(z) + g(z,x(z;\nu))), \quad \nu \ge 0\,.$$

We will then show (in $G$) uniform convergence of this sequence to a limit $x(z)$ which then satisfies the above equation. Moreover, we shall establish additional properties of $x(z)$, so that we may conclude with help of Proposition 3, Section 6.7, that $x(z)$ is the multi-sum of some formal power series (in inverse powers of $z$); this formal series then automatically satisfies (formally) the same differential equation, hence equals $\hat{x}(z)$, due to uniqueness of such formal solutions. In order to establish these additional properties, we shall have to consider not only one, but finitely many, sequences of functions and show their uniform convergence in corresponding regions, and we shall begin with defining these regions. To do this, we use the following notations:

For arbitrary $j$, $1 \leq j \leq q$, let $a_j$ be any eigenvalue of $A_j$; then $\tau$ is called a *Stokes' direction of level* $j$, iff $\text{Re} \left( a_j e^{i\tau k_j} \right) = 0$, i.e., iff the behavior of $\exp\{a_j z^{k_j}\}$ (as $z \to \infty$) changes at the ray $\arg z = \tau$. Obviously, $\tau$ is a Stokes' direction of level $j$, iff $\tau \pm \pi/k_j$ is so as well, so each halfopen interval of length $\pi/k_j$ contains the same (finite) number of such Stokes' directions; this number may be called $\mu_j$. The elements of the set of *all* Stokes' directions of level $j$ may be denoted by $\tau_\nu^{(j)}$, where $\nu$ runs through the set of integers, so that

$$\tau_\nu^{(j)} < \tau_{\nu+1}^{(j)}, \quad \tau_{\nu \pm \mu_j}^{(j)} = \tau_\nu^{(j)} \pm \pi/k_j,$$

for every $\nu$, and to make the enumeration unique, we may assume $\tau_{-1}^{(j)} < 0 \leq \tau_0^{(j)}$, but this will play no role in what follows.

Stokes' directions of different levels will not be related in any way, but it will be notationally convenient to assume that those of level $j$ are also Stokes' directions of level $j+1$; this can be made to hold by introducing "fictional" Stokes' directions of level $j+1$. For $\rho > 0$, sufficiently small $\varepsilon > 0$, $j = 1, \ldots, q$, and integer $\nu$, let

$$\begin{aligned} S_\nu^{(j)} &= S_\nu^{(j)}(\varepsilon, \rho) \\ &= \{z \mid |z| > \rho^{-1}, \tau_{\nu-\mu_j}^{(j)} + \varepsilon < \arg z < \tau_{\nu+1}^{(j)} - \varepsilon\}. \end{aligned}$$

Then (provided $\varepsilon$ is small enough) the family of $S_\nu^{(q)}$, for arbitrary integer $\nu$, covers the set $\{z \mid |z| > \rho^{-1}\}$ (on the Riemann surface of $\log z$) and for each $\nu$, finitely many sectors $S_\mu^{(q-1)}$ lie in $S_\nu^{(q)}$, and their union equals $S_\nu^{(q)}$, due to our introduction of fictional Stokes' directions. Generally, a sector $S_\nu^{(q-j+1)}$ will be the union of finitely many sectors $S_\mu^{(q-j)}$, each of which in turn is a union of finitely many sectors of level $q-j-1$, and so on. To indicate which sectors are contained in which ones, we shall give every sector of level $q-j+1$ several indices $\nu_1, \ldots, \nu_j$, namely so that for $j = 1, \ldots, q-1$:

$$S_{\nu_1, \ldots, \nu_j}^{(q-j+1)} = S_{\nu_j}^{(q-j+1)} \supset S_{\nu_1, \ldots, \nu_{j+1}}^{(q-j)} \quad (= S_{\nu_{j+1}}^{(q-j)}),$$

whenever $\nu_{j+1}$ runs through a set of consecutive integers, depending upon $\nu_j$; this set being denoted by $I_{\nu_j}^{(j)}$. Hence

$$S_{\nu_1, \ldots, \nu_j}^{(q-j+1)} = \bigcup_{\nu_{j+1} \in I_{\nu_j}^{(j)}} S_{\nu_1, \ldots, \nu_{j+1}}^{(q-j)}.$$

In order to proceed, we shall slightly shrink all the sectors $S_{\nu_1, \ldots, \nu_j}^{(q-j+1)}$, so that they become admissible in the following sense (to understand the meaning of this notion, you may wish to recall the exercises of the previous section): For $j = q + 1$, a region $G$ will be called *admissible of level* $q + 1$, iff for every $z \in G$ and every (real) $x \geq 1$ we have $xz \in G$ (so that in integrals from $z$ to $\infty$ we may choose the ray $\arg u = \arg z$, $|u| \geq |z|$, as path of integration). For $1 \leq j \leq q$, *admissible regions of level* $j$ will be such regions $G$ which (for some $\tilde{\varepsilon} > 0$, and $\rho$ as above) are, in the sense of the previous exercises, admissible with respect to $k_j$, arbitrary eigenvalues of $A_{j,\rho}$ and $\tilde{\varepsilon}$. While admissibility of level $q + 1$ would not present a problem, the above sectors will generally not be admissible of the other levels. So instead of sectors $S_\nu^{(q)}$, we define a region $G_\nu^{(q)}$ in the following manner: Make the change of variable $w = z^{k_q}$, mapping the sector $S_\nu^{(q)}$ onto a sector $\tilde{S}_\nu^{(q)}$ (on the Riemann surface of $\log w$); then define $\tilde{G}_\nu^{(q)}$ to be the subset of $\tilde{S}_\nu^{(q)}$ bounded by two halflines, parallel to the boundary rays of $\tilde{S}_\nu^{(q)}$ and tangent to the circle $|w| = \rho^{-k_q}$, and the corresponding part of this circle. Inverting the above mapping, we then obtain $G_\nu^{(q)}$ from $\tilde{G}_\nu^{(q)}$. The regions $G_\nu^{(q)}$ obtained in this manner are easily checked to be admissible with respect to both levels $q + 1$ and $q$.

The regions $G_\nu^{(q)}$ depend, in addition, upon $\rho$ and $\varepsilon$. If we momentarily allow $\varepsilon$ to take the value 0, then the union of all $G_\nu^{(q)}$ equals $\{z \mid |z| > \rho^{-1}\}$. For $\varepsilon > 0$, however small, this is not quite true, but instead the union misses certain "quasi-triangles" sitting upon the circle $|z| = \rho^{-1}$, but this will not be essential. It may, however, be helpful to observe for later use that these "quasi-triangles" have one large and two small inner angles which, as $\varepsilon \to 0$, tend to $\pi$ resp. 0 (in addition to the information that the whole "quasi-triangle" disappears for $\varepsilon \to 0$). For all these observations, it shall help to once-more employ the mapping $w = z^{k_q}$ and use its conformality (except for $z = 0$).

In case $q = 1$, we are done with our definition of admissible regions, while for $q \geq 2$ we continue as follows: For arbitrarily fixed $\nu$, the collection of sectors $S_{\nu,\mu}^{(q-1)}$, $\mu \in I_\nu^{(1)}$, obviously covers $G_\nu^{(q)}$. Both $G_\nu^{(q)}$ and $S_{\nu,\mu}^{(q-1)}$ depend upon the same number $\varepsilon > 0$, but we may in the definition of $S_{\nu,\mu}^{(q-1)}$ use any $\tilde{\varepsilon}$ with $0 < \tilde{\varepsilon} < \varepsilon$ and shall still have that the same $S_{\nu,\mu}^{(q-1)}$ are a covering of $G_\nu^{(q)}$. By means of the conformal map $w = z^{k_{q-1}}$, we transform $G_\nu^{(q)}$, resp. each $S_{\nu,\mu}^{(q-1)}$, into $\tilde{G}_\nu^{(q)}$, resp. $\tilde{S}_{\nu,\mu}^{(q-1)}$. The boundary curve of $\tilde{G}_\nu^{(q)}$ then has the asymptotes (as $w \to \infty$) $\arg w = k_{q-1}(\tau_\nu - \pi/k_q + \varepsilon)$ and $\arg w = k_{q-1}(\tau_{\nu+1} - \varepsilon)$, and furthermore, a tangent to the boundary curve will always remain in $\tilde{G}_\nu^{(q)}$ (due to the fact that $k_{q-1} > k_q$). For $\mu \in I_\nu^{(1)}$, except for the two extremal values, we define $\tilde{G}_{\nu,\mu}^{(q-1)}$ to be the subset of $\tilde{S}_{\nu,\mu}^{(q-1)}$ bounded by the two halflines, parallel to the boundary rays of $\tilde{S}_{\nu,\mu}^{(q-1)}$ and tangent to the boundary curve of $\tilde{G}_\nu^{(q)}$, and a corresponding part of the said boundary curve. For the maximal (minimal) value $\mu \in I_\nu^{(1)}$, the corresponding $\tilde{G}_{\nu,\mu}^{(q-1)}$ is defined similarly, but due to the fact that the left (right) boundary ray of $\tilde{S}_{\nu,\mu}^{(q-1)}$ will be outside of $\tilde{G}_\nu^{(q)}$, we draw only one halfline, parallel to the other boundary ray and tangent to the boundary curve of $\tilde{G}_\nu^{(q)}$. After inverting the mapping $w = z^{k_{q-1}}$, we have obtained the regions $G_{\nu,\mu}^{(q-1)}$, for $\mu \in I_\nu^{(1)}$. It then can be checked that these regions are admissible of level $q - 1$ (and $q + 1$), for whatever value $\tilde{\varepsilon} > 0$ (less than $\varepsilon$) we may have used in the construction. The union of $G_{\nu,\mu}^{(q-1)}$, $\mu \in I_\nu^{(1)}$, however, would be equal to $G_\nu^{(q)}$ only if we would allow $\tilde{\varepsilon} = 0$; whereas for $\tilde{\varepsilon} > 0$, the union differs from $G_\nu^{(q)}$ by a finite number of "quasi-triangles" sitting upon the boundary of $G_\nu^{(q)}$. But, quite analogous to the situation discussed above, the one large angle of them approaches $\pi$ as $\tilde{\varepsilon} \to 0$. This enables us, by taking $\tilde{\varepsilon}$ sufficiently small, to ensure that the union $\bigcup G_{\nu,\mu}^{(q-1)}$ (over $\mu \in I_\nu^{(1)}$) will be admissible of level $q$ (and $q+1$), and therefore we can replace $G_\nu^{(q)}$ by this union, which shall from now on be denoted by the same symbol $G_\nu^{(q)}$.

In case $q \geq 3$, we quite analogously define $G_{\nu_1,\nu_2,\nu_3}^{(q-2)}$, for every integer $\nu_1$, every $\nu_2 \in I_{\nu_1}^{(1)}$, and every $\nu_3 \in I_{\nu_2}^{(2)}$. In this construction, we use a third parameter $\hat{\varepsilon}$, with $0 < \hat{\varepsilon} < \tilde{\varepsilon}$, and again we have to remove, from each $G_{\nu_1,\nu_2}^{(q-1)}$ and each $G_{\nu_1}^{(q)}$, a finite number of "quasi-triangles". Taking $\hat{\varepsilon}$ sufficiently small, we can oncemore assure that the resulting new regions $G_{\nu_1,\nu_2}^{(q-1)}$, resp. $G_{\nu_1}^{(q)}$, are admissible of level $q-1$, resp. $q$ (and both are admissible of level $q+1$). Continuing this process, we complete, after finitely many steps, the definition of regions

$$G_{\nu_1,\dots,\nu_j}^{(q-j+1)},$$

being admissible of level $q-j$ (and $q+1$), and so that

$$G_{\nu_1,\dots,\nu_j}^{(q-j+1)} = \bigcup_{\nu_{j+1} \in I_{\nu_j}^{(j)}} G_{\nu_1,\dots,\nu_{j+1}}^{(q-j)},$$

for every $j = 1,\dots,q-1$.

**Remark.** It should be noted that for given indices $\nu_1,\dots,\nu_q$, with $\nu_{j+1} \in I_{\nu_j}^{(j)}$, $1 \leq j \leq q-1$, and $\tilde{\nu}_1,\dots,\tilde{\nu}_q$ with

$$\tilde{\nu}_j = \nu_j + 2k_j\mu_j, \quad 1 \leq j \leq q,$$

it follows that

$$\tilde{\nu}_{j+1} \in I_{\tilde{\nu}_j}^{(j)}, \quad 1 \leq j \leq q-1,$$

and

$$z \in G_{\tilde{\nu}_1,\dots,\tilde{\nu}_j}^{(q-j+1)} \quad \text{iff} \quad ze^{-2\pi i} \in G_{\nu_1,\dots,\nu_j}^{(q-j+1)}, \quad 1 \leq j \leq q.$$

It would suffice for the purposes we have in mind to restrict to finitely many regions covering one sheet of the Riemann surface, but it will be notationally more convenient not to do so.

## 8.4   Proper Iterations

In the previous section we defined regions

$$G_{\nu_1,\dots,\nu_j}^{(q-j-1)},$$

for $j = 1,\dots,q$, arbitrary integers $\nu_\tau$, and $\nu_{\tau+1} \in I_{\nu_\tau}^{(\tau)}$, $1 \leq \tau \leq j-1$. For short, tupels $(\nu_1,\dots,\nu_j)$ obeying these restrictions will be referred to as *admissible index tupels* (of level $q-j+1$). In addition to the parameters displayed, the regions $G_{\nu_1,\dots,\nu_j}^{(q-j+1)}$ also depend upon real parameters; one "*radius*" $\rho > 0$ and several

$\varepsilon$'s (one for each level). While the parameters $\varepsilon$ may be considered as fixed, the "radius" $\rho$ will play an essential role hereafter and should therefore be kept in mind.

In this section, we will show that for $\rho$ sufficiently large, and arbitrary admissible index tupels $(\nu_1, \ldots, \nu_q)$ (of level one), we can define a (unique) sequence of (vector) functions

$$x(z; \nu; \nu_1, \ldots, \nu_q), \quad \nu \geq 0,$$

having the following properties:

a) For $\nu = 0$ and every admissible index tupel $(\nu_1, \ldots, \nu_q)$ we have

$$x(z; 0; \nu_1, \ldots, \nu_q) \equiv 0.$$

b) For every admissible index tupel $(\nu_1, \ldots, \nu_q)$, and every $\nu \geq 1$, the function $x(z; \nu; \nu_1, \ldots, \nu_q)$ is analytic in $G^{(1)}_{\nu_1, \ldots, \nu_q}$. Moreover,

$$(A - z^{-K}\delta)x(z; \nu + 1; \nu_1, \ldots, \nu_q) = z^{-1}(b(z) + g(z, x(z; \nu; \nu_1, \ldots, \nu_q)))$$

for $\nu \geq 0$, $z \in G^{(1)}_{\nu_1, \ldots, \nu_q}$ and every admissible index tupel $(\nu_1, \ldots, \nu_q)$.

c) There exist constants $c > 0$, $M_1(\nu) > 0$ and $M_2(\nu) > 0$, independent of $z$ and $\nu_1, \ldots, \nu_q$ (and $c$ independent of $\nu$ as well), such that

$$\|x(z; \nu; \nu_1, \ldots, \nu_q)\|_\infty \leq M_1(\nu) \quad \text{in} \quad G^{(1)}_{\nu_1, \ldots, \nu_q},$$

for every $\nu \geq 0$ and every admissible index tupel $(\nu_1, \ldots, \nu_q)$, and if $(\tilde{\nu}_1, \ldots, \tilde{\nu}_q)$ also is admissible, then

$$\tilde{\nu}_\tau = \nu_\tau \quad \text{for} \quad 1 \leq \tau \leq q - j$$

(for some $j$) implies

$$\|x(z; \nu; \nu_1, \ldots, \nu_q) - x(z; \nu; \tilde{\nu}_1, \ldots, \tilde{\nu}_q)\|_\infty \leq M_2(\nu) \exp\{-c|z|^{k_j}\}$$

in $G^{(1)}_{\nu_1, \ldots, \nu_q} \cap G^{(1)}_{\tilde{\nu}_1, \ldots, \tilde{\nu}_q}$ (including the case of $j = q$, where the inequality is required to hold for arbitrary $(\tilde{\nu}_1, \ldots, \tilde{\nu}_q)$).

The construction of the regions $G^{(1)}_{\nu_1, \ldots, \nu_q}$ implies that for suitable $(\tilde{\nu}_1, \ldots, \tilde{\nu}_q)$ we have

$$z \in G^{(1)}_{\nu_1, \ldots, \nu_q} \quad \text{iff} \quad ze^{2\pi i} \in G^{(1)}_{\tilde{\nu}_1, \ldots, \tilde{\nu}_q},$$

and it may be seen, using the uniqueness of these functions, that this implies

$$x(z; \nu; \nu_1, \ldots, \nu_q) = x(ze^{2\pi i}; \nu; \tilde{\nu}_1, \ldots, \tilde{\nu}_q).$$

Consequently, we may restrict in a sense to finitely many admissible index tupels, but this does not play a role in most of what follows; on the contrary, it is notationally more convenient to allow arbitrary admissible $(\nu_1, \ldots, \nu_q)$.

Regarding the constant $c > 0$, we shall see that we may take any value $c$ provided that for arbitrary $j$, $1 \le j \le q$, and arbitrary eigenvalues $a_j$ of $A_j$, if

$$\exp\{a_j z^{k_j}\} \not\equiv 0 \quad \text{in} \quad G^{(1)}_{\nu_1,\ldots,\nu_q} \cap G^{(1)}_{\tilde{\nu}_1,\ldots,\tilde{\nu}_q}$$

(provided that the intersection is non-empty), then

$$|\exp\{a_j z^{k_j}\}| \le \exp\{-c|z|^{k_j}\} \quad \text{in} \quad G^{(1)}_{\nu_1,\ldots,\nu_q} \cap G_{\tilde{\nu}_1,\ldots,\tilde{\nu}_q} \, .$$

Existence of such a $c > 0$ may be seen to follow from the fact that the asymptotes of the boundary curve of arbitrary regions $G^{(1)}_{\nu_1,\ldots,\nu_q}$ are not equal to Stokes' rays of any level.

To show existence and uniqueness of these functions, assume for some $\nu \ge 0$

$$M_1(\nu) \le \rho_0/2 \, ,$$

which holds trivially in case $\nu = 0$, where we may take $M_1(0) = 0$. Assuming the parameter $\rho$ (in the construction of admissible regions) to satisfy

$$0 < \rho \le \rho_0/2 \, ,$$

we can define (for admissible $(\nu_1,\ldots,\nu_q)$) in $G^{(1)}_{\nu_1,\ldots,\nu_q}$

$$f(z;\nu;\nu_1,\ldots,\nu_q) := z^{-1}(b(z) + g(z, x(z;\nu;\nu_1,\ldots,\nu_q))) \, ,$$

and find for suitable $K > 0$, independent of $\rho, \nu, z$ and $\nu_1,\ldots,\nu_q$ :

$$\|f(z;\nu;\nu_1,\ldots,\nu_q)\|_\infty \le \rho K \quad \text{in} \quad G^{(1)}_{\nu_1,\ldots,\nu_q} \, .$$

Moreover, since $g(z,x)$, for $|z| \ge (\rho_0/2)^{-1}$ and $\|x\|_\infty \le \rho_0/2$, satisfies a Lipschitz condition with respect to $x$, with some Lipschitz constant $L > 0$, independent of $z$, we can easily show for $(\nu_1,\ldots,\nu_q)$ and $(\tilde{\nu}_1,\ldots,\tilde{\nu}_q)$ as in c) that in $G^{(1)}_{\nu_1,\ldots,\nu_q} \cap G^{(1)}_{\tilde{\nu}_1,\ldots,\tilde{\nu}_q}$

$$\|f(z;\nu;\nu_1,\ldots,\nu_q) - f(z;\nu;\tilde{\nu}_1,\ldots,\tilde{\nu}_q)\|_\infty \le \rho L M_2(\nu) \exp\{-c|z|^{k_j}\} \, .$$

To define the next "generation" of functions $x(z;\nu+1;\nu_1,\ldots,\nu_q)$, observe that in b) we may w.l.o.g. assume that the blocks $A_1,\ldots,A_{q+1}$ of the matrix $A$ are all in (lower triangular) Jordan canonical form. In this situation, the components of the vector $x(z;\nu+1;\nu_1,\ldots,\nu_q)$ can be recursively computed (and estimated), using the following

**Lemma 2.** *With* $k = (k_1,\ldots,k_q)$, $A = \text{diag}\,[A_1,\ldots,A_{q+1}]$ *and* $G^{(1)}_{\nu_1,\ldots,\nu_q}$ *as above, let* $r$ *be either zero or equal to* $k_j$, *for some (unique)* $j$, $1 \le j \le q$, *and let* $a \ne 0$ *denote an arbitrary eigenvalue of* $A_j$ *(with* $j = q+1$ *in case* $r = 0$ *). For each admissible index tupel* $(\nu_1,\ldots,\nu_q)$, *let* $f(z;\nu_1,\ldots,\nu_q)$ *be analytic in* $G^{(1)}_{\nu_1,\ldots,\nu_q}$ *and so that*

$$|f(z;\nu_1,\ldots,\nu_q)| \le M_1 \quad \text{in} \quad G^{(1)}_{\nu_1,\ldots,\nu_q} \, ,$$

with $M_1 > 0$, *independent of* $z$ *and* $\nu_1, \dots, \nu_q$. *Furthermore, for every admissible index set* $(\tilde{\nu}_1 \dots, \tilde{\nu}_q)$ *with* $\tilde{\nu}_\tau = \nu_\tau$ *for* $1 \leq \tau \leq q - \ell$, *let*

$$|f(z; \nu_1, \dots, \nu_q) - f(z; \tilde{\nu}_1, \dots, \tilde{\nu}_q)| \leq M_2 \exp\{-c|z|^{k_\ell}\}$$

*in* $G^{(1)}_{\nu_1, \dots, \nu_q} \cap G^{(1)}_{\tilde{\nu}_1, \dots, \tilde{\nu}_q}$. *Then for every admissible index tupel* $(\nu_1, \dots, \nu_q)$ *there exists a unique function* $g(z; \nu_1, \dots, \nu_q)$, *analytic in* $G^{(1)}_{\nu_1, \dots, \nu_q}$ *and satisfying*

$$(a - z^{-r}\delta)g(z; \nu_1, \dots, \nu_q) = f(z; \nu_1, \dots, \nu_q),$$

*so that for* $C > 0$, *independent of* $z, (\nu_1, \dots, \nu_q)$, $(\tilde{\nu}_1, \dots, \tilde{\nu}_q)$ *and* $f(z; \nu_1, \dots, \nu_q)$, *and also independent of* $\rho$, *we have*

$$|g(z; \nu_1, \dots, \nu_q)| \leq C(M_1 + M_2) \quad \text{in} \quad G^{(1)}_{\nu_1, \dots, \nu_q},$$

*and (in* $G^{(1)}_{\nu_1, \dots, \nu_q} \cap G^{(1)}_{\tilde{\nu}_1, \dots, \tilde{\nu}_q})$

$$|g(z; \nu_1, \dots, \nu_q) - g(z; \tilde{\nu}_1, \dots, \tilde{\nu}_q)| \leq C(M_1 + M_2) \exp\{-c|z|^{k_\ell}\}$$

*whenever* $\tilde{\nu}_\tau = \nu_\tau$ *for* $1 \leq \tau \leq q - \ell$.

**Proof.** In case $r = 0$, by assumption on the eigenvalues of $A_{q+1}$, we have $\text{Re}\, a > 0$, so that admissibility of $G^{(1)}_{\nu_1, \dots, \nu_q}$ of level $q + 1$ implies that

$$g(z; \nu_1, \dots, \nu_q) = z^{-a} \int_z^\infty w^{a-1} f(w; \nu_1, \dots, \nu_q) dw$$

(integrating along $\arg w = \arg z$) is the unique solution of the above differential equation which is bounded in $G^{(1)}_{\nu_1, \dots, \nu_q}$, for arbitrary admissible index tupel $(\nu_1, \dots, \nu_q)$, and the remaining properties can be easily shown. For the other cases, we apply Ex. 4, Section 8.2, with $G = G^{(j)}_{\nu_1, \dots, \nu_{q-j+1}}$ (for arbitrarily fixed admissible $(\nu_1, \dots, \nu_{q-j+1})$), and $G^{(1)}_{\nu_1, \dots, \nu_q}$ replacing the regions $G_\nu$ (for arbitrary $\nu_{q-j+2}, \dots, \nu_q$, so that $(\nu_1, \dots, \nu_q)$ is again admissible), and obtain existence and uniqueness of $g(z; \nu_1, \dots, \nu_q)$, and a corresponding $C > 0$ as stated, so that

$$|g(z; \nu_1, \dots, \nu_q)| \leq C(M_1 + M_2) \quad \text{in} \quad G^{(1)}_{\nu_1, \dots, \nu_q},$$

and (in $G^{(1)}_{\nu_1, \dots, \nu_q} \cap G^{(1)}_{\tilde{\nu}_1, \dots, \tilde{\nu}_q})$

$$|g(z; \nu_1, \dots, \nu_q) - g(z; \tilde{\nu}_1, \dots, \tilde{\nu}_q)| \leq C(M_1 + M_2) \exp\{-c|z|^{j-1}\}$$

whenever $\tilde{\nu}_\tau = \nu_\tau$ for $1 \leq \tau \leq q - j + 1$. To obtain the remaining estimates, distinguish the cases $1 \leq \ell \leq j - 2$, resp. $j \leq \ell \leq q$, (if they occur): In the first case, one easily sees that the better estimate for $f(z; \nu_1, \dots, \nu_q) - f(z; \tilde{\nu}_1, \dots, \tilde{\nu}_q)$ and the formula

$$[g(z; \nu_1, \dots, \nu_q) - g(z; \tilde{\nu}_1, \dots, \tilde{\nu}_q)] \exp\{-\frac{a}{r}z^r\}$$

$$= \int_r^\infty \exp\{-\frac{a}{r}w^r\}(f(w; \nu_1, \dots, \nu_q) - f(w; \tilde{\nu}_1, \dots, \tilde{\nu}_q))w^{r-1}dw$$

(following from Ex. 4, Section 8.2) imply the required estimate (possibly with enlarged $C > 0$). In the other case, it may so happen that in $G^{(1)}_{\nu_1,\ldots,\nu_q} \cap G^{(1)}_{\tilde\nu_1,\ldots,\tilde\nu_q}$ we find a ray along which $\exp\{-\frac{a}{r}z^r\} \cong 0$. If this is so, we can use the same path $\gamma(z)$ in the definition of both $g(z;\nu_1,\ldots,\nu_q)$ and $g(z;\tilde\nu_q,\ldots,\tilde\nu_q)$, and this path may stay completely within $G^{(1)}_{\nu_1,\ldots,\nu_q} \cap G^{(1)}_{\tilde\nu_1,\ldots,\tilde\nu_q}$. Using this, we easily obtain the desired estimate. On the other hand, if

$$\exp\{\frac{a}{r}z^r\} \cong 0 \quad \text{in} \quad G^{(1)}_{\nu_1,\ldots,\nu_q} \cap G^{(1)}_{\tilde\nu_1,\ldots,\tilde\nu_q} \, ,$$

then write

$$[g(z;\nu_1,\ldots,\nu_q) - g(z;\tilde\nu_1,\ldots,\tilde\nu_q)]\exp\{-\frac{a}{r}z^r\}$$

$$= c(z_0) + \int_{z_0}^z \exp\{-\frac{a}{r}w^r\}(f(w;\nu_1,\ldots,\nu_q) - f(w;\tilde\nu_q,\ldots,\tilde\nu_q))w^{r-1}dw \, .$$

with $z_0 \in G^{(1)}_{\nu_1,\ldots,\nu_q} \cap G^{(1)}_{\tilde\nu_1,\ldots,\tilde\nu_q}$, and

$$c(z_0) = [g(z_0;\nu_1,\ldots,\nu_q) - g(z_0;\tilde\nu_1,\ldots,\tilde\nu_q)]\exp\{-\frac{a}{r}z_0^r\} \, .$$

For suitably chosen $z_0$, we can guarantee

$$|\exp\{\frac{a}{r}(z^r - z_0^r)\}| \leq 1 \quad \text{in} \quad G^{(1)}_{\nu_1,\ldots,\nu_q} \cap G^{(1)}_{\tilde\nu_1,\ldots,\tilde\nu_q} \, ,$$

and then it is easy to estimate first $c(z_0)$ and then the integral and obtain the desired result.                                                                  □

Returning to our previous discussion, we find that, according to Lemma 2, we can uniquely define the functions $x(z;\nu+1;\nu_1,\ldots\nu_q)$, for every admissible index tupel $(\nu_1,\ldots,\nu_q)$, so that a) - c) again hold true, with

$$M_1(\nu+1) = M_2(\nu+1) = \rho C(K + LM_2(\nu)) \, ,$$

where $C > 0$ is a constant, *independent of* $\rho$. Choosing $\rho$ sufficiently small, we can ensure that

$$M_1(\nu+1) \leq \rho_0/2 \, .$$

In this manner, we can inductively prove existence and uniqueness of the sequence of functions $x(z;\nu;\nu_1,\ldots,\nu_q)$, $\nu \geq 0$, for every admissible index vector $(\nu_1,\ldots,\nu_q)$.

## 8.5   The Convergence Proof

In this section we shall show uniform convergence of the sequences $x(z;\nu;\nu_1,\ldots,\nu_q)$, $\nu \geq 0$, for each admissible index tupel $(\nu_1,\ldots,\nu_q)$, on the region $G^{(1)}_{\nu_1,\ldots,\nu_q}$, and this then leads to a proof of the following

**Theorem 1.** *Let a normalized system of non-linear differential equations, as stated in Section 8.3, be given, and let regions* $G^{(1)}_{\nu_1,\ldots,\nu_q}$, *for every admissible index tupel* $(\nu_1,\ldots,\nu_q)$, *be as described in the same section. Then there exist functions*

$$x(z;\nu_1,\ldots,\nu_q)$$

*(for* $(\nu_1,\ldots,\nu_q)$ *as above), having the following properties:*

a) *For every admissible index tupel* $(\nu_1,\ldots,\nu_q)$, *the function is analytic and bounded in* $G^{(1)}_{\nu_1,\ldots,\nu_q}$, *and satisfies*

$$(A - z^{-K}\delta)x(z;\nu_1,\ldots,\nu_q) = z^{-1}(b(z) + g(z,x(z;\nu_1,\ldots,\nu_q)))\,.$$

b) *For any two admissible index tupels* $(\nu_1,\ldots,\nu_q)$ *and* $(\tilde\nu_1,\ldots,\tilde\nu_q)$ *with*

$$\tilde\nu_\tau = \nu_\tau \quad \text{for} \quad 1 \le \tau \le q-j$$

*(for some* $j$, $1 \le j \le q$*), and for some* $c > 0$, *the function*

$$\|x(z;\nu_1,\ldots,\nu_q) - x(z;\tilde\nu_1,\ldots,\tilde\nu_q)\| \exp\{c|z|^{k_j}\}$$

*is bounded in* $G^{(1)}_{\nu_1,\ldots,\nu_q} \cap G^{(1)}_{\tilde\nu_1,\ldots,\tilde\nu_q}$.

**Proof.** As indicated above, we only have to prove uniform convergence of $x(z;\nu;\nu_1,\ldots,\nu_q)$, as $\nu \to \infty$ (for $z \in G^{(1)}_{\nu_1,\ldots,\nu_q}$), for every admissible index tupel $(\nu_1,\ldots,\nu_q)$. To do so, let us abbreviate

$$d(z;\nu;\nu_1,\ldots,\nu_q) = x(z;\nu+1;\nu_1,\ldots,\nu_q) - x(z;\nu;\nu_1,\ldots,\nu_q)\,,$$

$$h(z;\nu;\nu_1,\ldots,\nu_q) = z^{-1}(g(z,x(z;\nu+1;\nu_1,\ldots,\nu_q)) - g(z,x(z;\nu;\nu_1,\ldots,\nu_q)))$$

for $\nu \ge 0$, $z \in G^{(1)}_{\nu_1,\ldots,\nu_q}$. For some $\nu \ge 0$, assume existence of $D_1(\nu), D_2(\nu) \ge 0$, so that for arbitrary admissible index tupels $(\nu_1,\ldots,\nu_q)$ and $(\tilde\nu_1,\ldots,\tilde\nu_q)$ (and $c > 0$ as in the previous section) the following holds:

$$\|d(z;\nu;\nu_1,\ldots,\nu_q)\|_\infty \le D_1(\nu) \quad \text{in} \quad G^{(1)}_{\nu_1,\ldots,\nu_q}\,,$$

and if

$$\tilde\nu_\tau = \nu_\tau \quad \text{for} \quad 1 \le \tau \le q-j$$

(for some $j$, $1 \le j \le q$), then

$$\|d(z;\nu;\nu_1,\ldots,\nu_q) - d(z;\nu;\tilde\nu_1,\ldots,\tilde\nu_q)\|_\infty \le D_2(\nu)\exp\{-c|z|^{k_j}\}$$

in $G^{(1)}_{\nu_1,\ldots,\nu_q} \cap G^{(1)}_{\tilde\nu_1,\ldots,\tilde\nu_q}$. This is certainly correct for $\nu = 0$, and then implies

$$\|h(z;\nu;\nu_1,\ldots,\nu_q)\|_\infty \le \rho L D_1(\nu) \quad \text{in} \quad G^{(1)}_{\nu_1,\ldots,\nu_q}\,,$$

and (with help of the exercise at the end of this section)

$$\|h(z;\nu;\nu_1,\ldots,\nu_q) - h(z;\nu;\tilde\nu_1,\ldots,\tilde\nu_q)\| \le \rho M(M_2(\nu)2D_1(\nu) + D_2(\nu))\exp\{-c|z|^{k_j}\}$$

in $G^{(1)}_{\nu_1,\ldots,\nu_q} \cap G^{(1)}_{\tilde{\nu}_1,\ldots,\tilde{\nu}_q}$, for some $M > 0$, independent of $\nu, z$, and $(\nu_1,\ldots,\nu_q)$, $(\tilde{\nu}_1,\ldots,\tilde{\nu}_q)$, and $M_2(\nu)$ $(\leq \rho_0/2)$ as in the previous section. From Lemma 2 we then conclude

$$D_1(\nu+1) = D_2(\nu+1) \leq \rho C(LD_1(\nu) + M(M_2(\nu)2D_1(\nu) + D_2(\nu)))\,.$$

For sufficiently small $\rho \geq 0$, this implies

$$D_1(\nu) = D_2(\nu) \leq \alpha^{\nu-1}D\,, \quad D = D_1(1) = D_2(1)\,,$$

with $\alpha \in (0,1)$, and this clearly gives uniform convergence of $x(z;\nu;\nu_1,\ldots,\nu_q)$, as $\nu \to \infty$. □

**Remark.** Since each $x(z;\nu_1,\ldots,\nu_q)$ is the limit of the sequence $x(z;\nu;\nu_1,\ldots,\nu_q)$ for $\nu \to \infty$, we conclude for $(\tilde{\nu}_1,\ldots,\tilde{\nu}_q)$ with $\tilde{\nu}_j = \nu_j + 2k_j\mu_j$, $1 \leq j \leq q$, that

$$x(z;\nu_1,\ldots,\nu_q) = x(ze^{2\pi i};\tilde{\nu}_1,\ldots,\tilde{\nu}_q) \quad \text{in} \quad G^{(1)}_{\nu_1,\ldots,\nu_q}\,.$$

As an application of Theorem 1, we now obtain the promised result on multisummability of the formal solution $\hat{x}(z)$ of our normalized system of differential equations. As was pointed out in Section 8.1, this then implies multisummability of *any* formal solution to arbitrary (non-linear) systems of ODE.

**Theorem 2.** *Let a normalized system of non-linear differential equations, as stated in Section 8.3, be given, and let $\hat{x}(z)$ be its (unique) formal solution (in form of a power series in $z^{-1}$). Then*

$$\hat{x}(z^{-1}) \in \mathbb{C}\{z\}_k^n\,.$$

**Proof.** Let $d = (d_1,\ldots,d_q)$ be a multi-direction, admissible with respect to $k$, and such that for each $j$, $1 \leq j \leq q$, no eigenvalue of $A_j$ or $-A_j$ lie on the ray

$$\arg z = k_j d_j\,.$$

Modulo $2\pi$, this condition excludes finitely many multi-directions, so it suffices to show

$$\hat{x}(z^{-1}) \in \mathbb{C}\{z\}_{k,d}^n\,,$$

for every such $d$. To do this, recall the definition of regions to see that there is exactly one admissible index tupel $(\nu_1,\ldots,\nu_q)$ for which the region $G^{(q-j+1)}_{\nu_1,\ldots,\nu_j}$ contains a sector with bisecting direction $d_{q-j+1}$ and opening more than $\pi/k_{q-j+1}$, for every $j = 1,\ldots,q$. So for $\varepsilon > 0$ and $r > 0$, sufficiently small (and intervals $I_1,\ldots,I_q$ as in Section 6.7) we have $S_\delta := S(\delta,\varepsilon,r) \subset G^{(q-j+1)}_{\nu_1,\ldots,\nu_j}$, for every $\delta \in I_{q-j+1}$. Hence for $\delta \in I_1$, we may define

$$f(z;\delta) = x(z;\nu_1,\ldots,\nu_q) \quad \text{in} \quad S_\delta\,,$$

while for $j = 1,\ldots,q-1$ and $\delta \in I_{q-j+1}\backslash I_{q-j}$, we can find an admissible index tupel $(\tilde{\nu}_1,\ldots,\tilde{\nu}_q)$, with

$$\tilde{\nu}_\tau = \nu_\tau\,, \quad 1 \leq \tau \leq j\,,$$

so that $S_\delta \subset G^{(1)}_{\tilde{\nu}_1,\ldots,\tilde{\nu}_q}$, and we then define

$$f(z;\delta) = x(z;\tilde{\nu}_1,\ldots,\tilde{\nu}_q) \quad \text{in} \quad S_\delta .$$

Finally, for $\delta \notin I_q$, we take any $(\tilde{\nu}_1,\ldots,\tilde{\nu}_q)$ with $S_\delta \subset G^{(1)}_{\tilde{\nu}_1,\ldots,\tilde{\nu}_q}$ and define $f(z;\delta)$ in the same way. Having done so, one finds from Theorem 2, Section 7.3, that for $\delta \in I_1$, the function $f(z^{-1};\delta)$ is the multisum (in the multi-direction $d$) of some formal power series $\hat{f}(z^{-1})$. This series then has to be a formal solution to our system of ODE, and by uniqueness of such a formal solution we find $\hat{f} = \hat{x}$. $\quad\square$

**Exercise.**

Let a vector $g(z,x)$, $x = (x_1,\ldots,x_n)^T$, be analytic in every variable, for

$$|z| > \rho_0^{-1}, \quad \|x\|_\infty < \rho_0$$

(including $z = \infty$). For

$$|z| \geq 2\rho_0^{-1}, \quad \|x_j\|_\infty, \|y_j\|_\infty \leq \rho_0/2 \quad 1 \leq j \leq 2 ,$$

show existence of a constant $M$ (independent of $z, x_1, x_2, y_1, y_2$), so that

$$\|g(z,x_1) - g(z,y_1) - (g(z,x_2) - g(z,y_2))\|_\infty$$
$$\leq M\{\|x_1 - y_1\|_\infty(\|x_1 - x_2\|_\infty + \|y_1 - y_2\|_\infty) + \|x_1 - y_1 - (x_2 - y_2)\|_\infty\} .$$

*Hint.* Expanding $g(z,x)$, for fixed $z$, into a power series about $y$ with $\|y\|_\infty < \rho_0$, conclude existence of a matrix $H(z,x,y)$, analytic in all variables, so that

$$g(z,x) = g(z,y) + H(z,x,y)(x - y) .$$

Use this to conclude

$$g(z,x_1) - g(z,y_1) - (g(z,x_2) - g(z,y_2))$$
$$= (H(z,x_1,y_1) - H(z,x_2,y_2))(x_1 - y_1)$$
$$+ H(z,x_2,y_2)(x_1 - y_1 - (x_2 - y_2)) ,$$

and estimate this formula.

# References

[Ba 1]    W. Balser, A different characterization of multisummable power series, Analysis 12 (1992) 57–65.

[Ba 2]    W. Balser, Summation of formal power series through iterated Laplace integrals, Math. Scand. 70 (1992) 161–171.

[Ba 3]    W. Balser, Einige Beiträge zur Invariantentheorie meromorpher Differentialgleichungen, Habilitationsschrift, Ulm (1978).

[Ba 4]    W. Balser, Solutions of first level of meromorphic differential equations, Proc. Edinburgh Math. Soc. 25 (1982) 183–207.

[Ba 5]    W. Balser, Growth estimates for the coefficients of generalized formal solutions, and representation of solutions using Laplace integrals and factorial series. Hiroshima Math. J. 12 (1982) 11–42.

[Ba 6]    W. Balser, A constructive existence proof for first level formal solutions of meromorphic differential equations, Hiroshima Math. J. 15 (1985) 411–427.

[Ba 7]    W. Balser, Addendum to my paper: A different characterization of multisummable power series, Analysis 13 (1993) 317–319.

[BBRS]    W. Balser, B.L.J. Braaksma, J.-P. Ramis, and Y. Sibuya, Multisummability of formal power series solutions of linear ordinary differential equations, Asymptotic Analysis 5 (1991) 27–45.

[Bi]      G.D. Birkhoff, Singular points of ordinary linear differential equations, Trans. AMS 10 (1909) 436–470.

[BJL]     W. Balser, W.B. Jurkat, and D.A. Lutz, Characterization of first level formal solutions by means of the growth of their coefficients, J. Diff. Equations 51 (1984) 48–77.

[BT]      W. Balser and A. Tovbis, Multisummability of iterated integrals, Asympt. Anal. 7 (1993) 121–127.

[Bo]      R.P. Boas, Entire Functions, Acad. Press 1954.

[Br 1]   **B.L.J. Braaksma**, Multisummability of formal power series solutions of
         nonlinear meromorphic differential equations, Ann. Inst. Fourier, Greno-
         ble 42 (1992) 517–540.

[Br 2]   **B.L.J. Braaksma**, Laplace integrals in singular differential and difference
         equations, in: Proc. Conf. Ordinary and Partial Differential Equations,
         Dundee 1978, Lecture Notes in Math. 827 (1980) 25–53.

[Br 3]   **B.L.J. Braaksma**, Multisummability and Stokes multipliers of linear
         meromorphic differential equations, J. Diff. Eq. 92 (1991) 45–75.

[BV]     **D.G. Babbitt and V.S. Varadarajan**, Some remarks on the asymptotic
         existence theorem for meromorphic differential equations, J. Fac. Sci.
         Univ. Tokyo 36 (1989) 247–262.

[CL]     **E.A. Coddington and N. Levinson**, Theory of Ordinary Differential
         Equations, McGraw-Hill, 1955.

[Ec 1]   **J. Ecalle**, Les fonctions résurgents I–III, Publ. Math. d'Orsay, 1981 and
         1985.

[Ec 2]   **J. Ecalle**, Introduction à l'accélération et à ses applications, Traveaux en
         Cours, Hermann 1993.

[Er]     **A. Erdélyi**, Asymptotic Expansions, Dover 1956.

[Hi]     **E. Hille**, Ordinary Differential Equations in the Complex Domain, Wiley
         1976.

[HI]     **M. Hukuhara and M. Iwano**, Étude de la convergence des solutions
         formelles d'un système différentiel ordinaire linéaire, Funk. Ekvac. 2
         (1959) 1–18.

[Ho 1]   **J. Horn**, Fakultätenreihen in der Theorie der linearen Differentialgle-
         ichungen, Math. Ann. 71 (1912) 510–532.

[Ho 2]   **J. Horn**, Integration linearer Differentialgleichungen durch Laplacesche
         Integrale und Fakultätenreihen, Jahresber. DMV 24 (1915) 309–329, und
         25 (1916) 74–83.

[Ho 3]   **J. Horn**, Integration linearer Differentialgleichungen durch Laplacesche
         Integrale I, II, Math. Z. 49 (1944) 339–350 und 684–701.

[Hu 1]   **M. Hukuhara**, Intégration formelle d'un système d'équations
         différentielles non linéaires dans le voisinages d'un point singulier, Ann.
         Mat. P. Appl. Bologna 19 (1940) 35–44.

[Hu 2]   **M. Hukuhara**, Sur les points singuliers des équations différentielles
         linéaires III, Mém. Fac. Sci. Kyushu 2 (1942) 35–44.

[In]  E.L. Ince, Ordinary Differential Equations, Dover 1956.

[Ju 1]  W.B. Jurkat, Meromorphe Differentialgleichungen, Lecture Notes in Math. 637 (1978), Springer Verlag.

[Ju 2]  W.B. Jurkat, Summability of asymptotic power series, Asympt. Anal. 7 (1993) 239–250.

[MgR]  B. Malgrange and J.-P. Ramis, Fonctions multisommable, Ann. Inst. Fourier, Grenoble 41 (1991) 1–16.

[MR]  J. Martinet and J.-P. Ramis, Elementary acceleration and multi-summability, Preprint I.R.M.A. Strasbourg, 428/P-241, 1990, Ann. Inst. Henri Poincaré, Physique Theorique 54 (1991) 331–401.

[Ne]  F. Nevanlinna, Zur Theorie der asymptotischen Potenzreihen, Ann. Acad. Scient. Fennicae, Helsinki 1918.

[Ol]  F.W.J. Olver, Asymptotics and Special Functions, Acad. Press, 1974.

[Ra 1]  J.-P. Ramis, Les séries $k$-sommable et leurs applications, in: Springer Lecture Notes in Physics 126 (1980) 178–199.

[Ra 2]  J.-P. Ramis, Dévissage Gevrey, Asterisque 59/60 (1978) 173–204.

[RS 1]  J.-P. Ramis and Y. Sibuya, Hukuhara domains and fundamental existence and uniqueness theorems for asymptotic solutions of Gevrey type, Asymptotic Analysis 2 (1989) 39–94.

[RS 2]  J.-P. Ramis and Y. Sibuya, A theorem concerning multisummability of formal solutions of non linear meromorphic differential equations, Manuscript 1993.

[SG]  G. Sansone and J. Gerretsen, Lectures on the Theory of Functions of a Complex Variable, P. Noordhoff, 1960.

[Si]  Y. Sibuya, Linear Differential Equations in the Complex Domain: Problems of Analytic Continuation, Transl. Math. Monographs AMS 82 (1990).

[Ti]  E.C. Titchmarsh, The Theory of Functions, Oxford 1939.

[Tr]  W.J. Trjitzinsky, Laplace integrals and factorial series in the theory of linear differential and difference equations, Trans AMS 37 (1935) 80–146.

[Tu]  H.L. Turrittin, Convergent solutions of ordinary linear homogeneous differential equations in the neighborhood of an irregular singular point, Acta Math. 93 (1955) 27–66.

[Wa]    **W. Wasow**, Asymptotic Expansions of Ordinary Differential Equations, Dover (1987).

[Wi]    **D.V. Widder**, The Laplace Transform, Princeton 1941.

[Wt 1]  **G.N. Watson**, A theory of asymptotic series, Trans Royal Soc. London, Ser. A 211 (1911) 279–313.

[Wt 2]  **G.N. Watson**, The transformation of an asymptotic series into a convergent series of inverse factorials, Rend. circ. Pal. 34 (1912) 41–88.

# Index

# List of Symbols